MARINE
OF THE MEDITERRANEAN

MARINE LIFE
OF THE MEDITERRANEAN

LAWSON WOOD

BLOOMSBURY WILDLIFE
LONDON · OXFORD · NEW YORK · NEW DELHI · SYDNEY

BLOOMSBURY WILDLIFE
Bloomsbury Publishing Plc
50 Bedford Square, London, WC1B 3DP, UK
29 Earlsfort Terrace, Dublin 2, Ireland

BLOOMSBURY, BLOOMSBURY WILDLIFE and the Diana logo
are trademarks of Bloomsbury Publishing Plc

This edition published in the United Kingdom 2024

A catalogue record for this book is available from the
British Library.

Library of Congress Cataloguing-in-Publication
data has been applied for.

ISBN: PB: 978-1-399-41170-7
ePub: 978-1-399-41168-4
ePDF: 978-1-399-41167-7

10 9 8 7 6 5 4 3 2 1

Designed by Austin Taylor
Map by Brian Southern
Printed and bound in Turkey by Elma Basim

100%
From well-
managed forests
FSC
www.fsc.org   FSC® C164814

To find out more about our authors and
books visit www.bloomsbury.com and sign
up for our newsletters.

CONTENTS

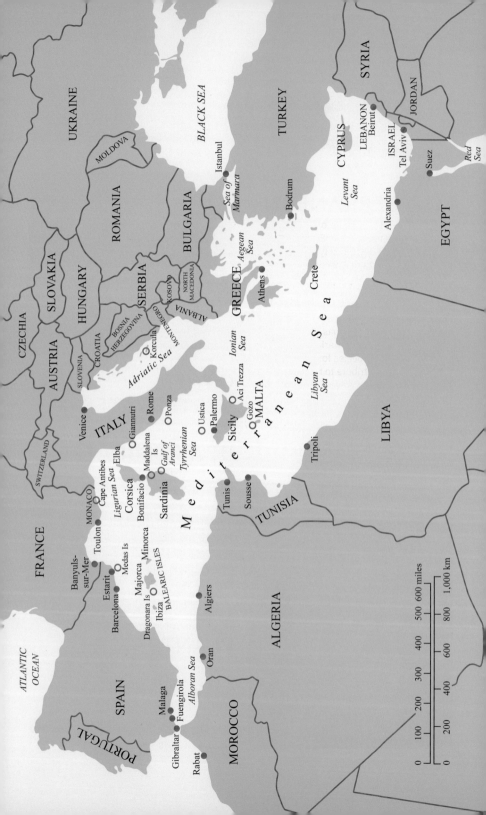

THE MEDITERRANEAN

KNOWN AS THE 'Cradle of Civilisation', the Mediterranean Sea is set in a massive flooded depression in the Earth's crust, formed over millennia. This vast expanse of water is almost completely landlocked, stretching from the Strait of Gibraltar in the west to the coasts of Turkey, Syria and Israel in the East. It is also linked to the Black Sea in the north-east, through the Sea of Marmara and the Dardanelles.

The Mediterranean both links and divides three continents – Europe, Africa and Asia. The varied depths of this basin also support varying communities of marine life, and although all of the species found in the Mediterranean contribute to the whole, we shall only concern ourselves with what we can see, practically and with the least effort, whether we are swimming, snorkelling, scuba diving, or even fishing and sailing.

With water temperatures that never drop below 10°C, the Mediterranean has remained fairly isolated over the years and has evolved a distinct ecosystem only troubled by human intrusion. The region is complex

▼ The near shore of the Mediterranean is dominated by seagrass, algae and small sponges.

and diverse in its geography, history and climate, with much of what we know and see only coming from the northern shores and around many of the islands.

Known for its very singular climate, the Mediterranean's present shores would have been recognisable to us five million years ago. However, the sea took early form more than 20 million years ago when the western shores were still closed and the eastern end was still linked to the vast, primeval Tethys Sea. As the years passed, Africa slipped anticlockwise, Arabia split from Africa to form the Red Sea, and the eastern stretches of the Mediterranean also became landlocked. Whilst this continental movement was progressing, the land areas were gradually 'crumpled' up, forming a ring of mountains which now comprise the Sierra Nevada, the Alps, the Atlas Mountains, the Dinaric Alps, the Rhodope Mountains, the Akhdar Heights and the Taurus Mountains. Over the subsequent two or three million years, the now-landlocked Mediterranean filled and evaporated several times as the Earth's crust shifted, often leaving gigantic saltmarshes and mud pools, where long-extinct animals once roamed.

As far as can be agreed between scientists, about five million years ago there was a massive cataclysmic earthquake which opened up the Gibraltar sill, and the Atlantic Ocean poured into the Mediterranean basin, taking almost a century to fill. Today, this connection to the Atlantic Ocean is still very evident and water pours into the Mediterranean at approximately 4km/h (2.5mph) in a layer from 75–300m (250–1,000ft) deep. This sea water gradually evaporates and the heavier, more salty water sinks and eventually flows back out through the Strait of Gibraltar, taking around 80 years to replenish the water column. This has had several knock-on effects, one of which is the reduced amount of plankton to be found in the Mediterranean. What plankton there is, quite often flows back out into the Atlantic in the deeper ocean currents.

This ultimately produces less marine life, but also makes for clearer water in general. (Scientists generally agree that there are currently around 17,000 named species of organisms in the Mediterranean, of which 26 per cent are micro-organisms – with more arriving each year from the Red Sea, resulting in more than 1,000 invasive species so far.) Invasive species from the Red Sea through the Suez Canal are known as *Lessepsian* species. As always, there are notable exceptions and wherever there is a large river run-off there will be periodic algal blooms which reduce water clarity. Rough seas in shallow areas also contribute to poor visibility, but in general one should expect underwater visibility of about 30–50m (100–165ft).

Endemic species

Natural, evolving endemism in the Mediterranean began around 5 million years ago when the western Gibraltar sill was breached by a cataclysmic earthquake, and this almost landlocked basin gradually filled over the next 100 years or so. Due to the strength of the currents, many marine species were carried along and then virtually trapped in this enclosed sea. They gradually adapted to their new environment and subtly changed over the generations, through natural selection. Many are very closely related to their neighbours in the Atlantic and Indian Oceans, and for some you will need scientific examination, rather than wishful thinking, to separate the species. With the initial opening of the Suez Canal in 1869, there was a very slow but steady influx of marine species flowing north into the Mediterranean from the Indian Ocean via the Red Sea. However, since the dredging and widening of the canal in 2015, this influx of foreign species has increased dramatically.

Suez Canal

Canal works to link the Nile and its delta to the Bitter Lakes and the Red Sea were first started during the reign of the Pharaohs; specifically Pharaoh Senustret III of the Twelfth Dynasty (1897–1839 BC). However, this work was abandoned when he discovered that the sea level of the Red Sea was higher than the Nile. Further works were undertaken by the Persian King Darius I, who managed to connect Lake Timsah with the Great Bitter Lake. Ptolemy II extended this by excavating a trench 30m (100ft) wide by 9m (30ft) deep and connected the lakes further. Pharaoh Hatshepsut was said to have finally connected the Red Sea with the Nile Delta (and thence the Mediterranean) by the thirteenth century BC.

Napoleon Bonaparte also tried to extend and deepen these ancient canal works, but this project was abandoned when it was discovered that his engineers' calculations were wrong.

It wasn't until the mid 1800s, when Ferdinand de Lesseps convinced the Egyptian viceroy Said Pasha to support the project, that construction began. The Suez Canal (as we know it today) now connects Port Said in the Red Sea to Alexandria on the Mediterranean, and is 193.30km (120.11 miles) long. The canal was officially opened on 7 November 1869. After the canal was closed from 1967 to 1975 because of the Six-Day War, it failed to regain its trading supremacy until it was dredged and widened in 2015.

The current in the canal flows north, so waters (along with their marine life) now flow 'downwards' from the Red Sea into the Mediterranean. This has changed the ecosystem of the eastern Mediterranean forever.

A SEA OF SEAS

Although not truly one sea, but several joined together, the individual seas that form the Mediterranean are defined by the surrounding land masses formed when the Earth's crust collided. There are two principal areas, known as the Eastern and Western Mediterranean and roughly separated by the mid-ocean ridge which runs from Italy to the African coast. The Alborán Sea is found in the west, connecting to the Algerian Sea or Balearic Basin between Algeria, Sardinia, Spain and France. The Gulf of Lyons and the Ligurian Sea stretch along the southern coast of France, and the Tyrrhenian Sea is surrounded by Corsica and Sardinia to the west, Italy to the east and Sicily to the south.

The two most famous of the Mediterranean's seas are the Adriatic and the Aegean; it is on these seas that the history of humanity spread from the old to the new worlds. The Gulf of Trieste, in the northern area of the Adriatic Sea, is formed

▶ Corfu, a Greek island, is a popular tourist destination.

with Italy on its west, Croatia and Albania to its east. Further south the Ionian Sea separates the southern coasts of Italy and Sicily, with the Maltese islands to the west and Greece and its many islands to the east. Separating Greece and Turkey is the Aegean Sea, which is connected to the Black Sea in the north-east by the Dardanelles, Sea of Marmara and the Bosporus. In the eastern reaches can be found the Levant Sea, bordered by Turkey and the islands of Crete and Cyprus to the north and Syria, Lebanon and Israel to the east, with Egypt and Libya to the south. The Ionian Sea and the Levant Sea are roughly divided by a shallow stretch of water that is sometimes referred to as the Libyan Sea. The Ionian basin is the deepest part of the Mediterranean, at 5,121m (17,000ft). In Egypt, of course, can be found the artificial Suez Canal, opened in 1869, which links the Mediterranean with the Red Sea.

HUMAN IMPACT

As the Mediterranean is bounded by mountains, human impact has been concentrated mainly on the narrow coastal plain, some 20–30km (13–20 miles) wide. In some areas of southern France and northern Italy, people have literally carved a space from the mountains themselves. With the one notable exception along the North African coast, virtually all of the Mediterranean's population can be found along this narrow coastal strip and its associated islands. Much of the coastal plain was dominated by mosquito-infested swamps, and history tells of the canal and irrigation systems which drained the swamps and irrigated the land for cultivation, thereby virtually eradicating malaria. Much of the original forest has been cut down, leaving a rather sparse scrubby landscape.

The Mediterranean basin and surrounding countries are home to some 529 million people; approximately 175 million of them live around its shores. Economic and industrial growth has resulted in vastly increased pollution levels. The need to feed this increasing population, and its annual tourist trade of more than 300 million visitors, has put further strain on the sea and its marine life. Much of the original coastline has been irrevocably changed by human population expansion, and associated development, including yacht marinas, harbours

◀ This former ferry was deliberately sunk off Gozo as an artificial reef and diver attraction.

An alien invasion

There are many ways for alien, invasive species to reach the Mediterranean, but principally it is through direct human activity, with the main culprit being global trade with commercial vessels. There is great concern over alien invaders accidentally brought to the Mediterranean from other seas, whether it be by oceanic shipping discharging their ballast tanks, or carrying the organisms attached to their hulls. Aquaculture has also introduced hundreds of alien species attached to oyster spats. More than 46 per cent of non-native species have arrived this way. The natural influx of plankton and other species from the greater Atlantic has obviously been the largest influencer on the ecosystem over many thousands of years, but now with the expansion of the Suez Canal, literally hundreds and hundreds of Lessepsian species are now arriving from further east, and have quite literally forever changed the eastern Mediterranean's ecology. The United Nations Environment Programme Mediterranean Action Plan indicated that the Mediterranean is warming 20 per cent faster than any other sea. This makes it ever more hospitable for cross-over species from the Red Sea.

The Red Sea can of course now be a receiver of tropical species, with the Suez Canal acting as a 'revolving door' as species from both major oceans intermingle at the confluences of the

▲ The Rapa Whelk is originally from Asia and is now so successful in the Mediterranean, it is now commercially harvested.

canal and greater Red Sea. This book describes the more significant migrant species which are spreading through the Mediterranean.

One example is the Rapa Whelk (*Rapana venosa*), thought to have been ejected in ballast water from a cargo ship in the Black Sea back in 1947. This species had no natural enemies in its new home, and quickly expanded its population. There is now a commercial industry in the Mediterranean that is based around harvesting this originally Pacific snail, and exporting it back to the Asian community.

Are there other plus sides to this ecological shift? Some scientists now consider that the Mediterranean is being 're-wilded' but with a different range of species. There will eventually be a natural order, and a full ecosystem of predators and prey, but for now, scuba divers and snorkellers can enjoy a bounty of bright and colourful subtropical fish to view in their underwater encounters. Conservation efforts from now on are focused on gaining more protection for coastal marine sites, as well as wider expanses of open seas, from mass fishing practices.

and housing developments. Small countries such as Monaco have now lost some 75 per cent of their original coastline (however, all of Monaco's coastline – artificial and original – is protected). The Mediterranean Sea is undoubtedly one of the world's most threatened seas, owing solely to the increased pressure on its natural resources. The primary threat is pollution from homes, industry and intensive agricultural and aquaculture methods. Fortunately, the fresh seawater influx from the Atlantic has managed to mitigate much of this, at least for the time being.

The largest specific single threat now (notwithstanding the effects of climate change) is the increase of warm water, nutrients and marine species through the Suez Canal. The canal was dredged and widened in 2015, thus removing the natural high salinity basin of the Bitter Lakes, which up until 2015 almost stopped the major influx of marine life from the Red Sea into the Mediterranean. Now the number of new species being recorded in the Mediterranean has multiplied exponentially. For example, there are now records of Common Lionfish (*Pterois volitans*) from around Cyprus, the Greek Islands and even Malta. There is a similar invasion of Common Lionfish in the Caribbean. It is thought that this was the result of hurricane-damaged aquaria which released their occupants into the sea. Curiously, whilst the influx of this species and other reef fish may have a future devastating effect on the Mediterranean ecosystem, scuba divers and snorkellers are actually enjoying seeing this much wider variety of colourful fish around the shores of the Mediterranean.

CARPET ALGAE

Even though the major problems of pollution are now being taken on board by the Mediterranean's surrounding countries, the impact on the local sea environment is actually very small considering the size of the Mediterranean Sea. Small areas such as around Venice and the northern Adriatic, sections of the Greek coast and parts of Tunisia are under increased threat owing to tourism demands, but this is usually on a temporary seasonal basis. The marine life does regenerate in these areas with seasonal changes. However, as if this form of human impact were not enough in altering and destroying the coastline, there is now the continuing threat of carpet algae. In 1984, a small patch of an alga called *Caulerpa taxifolia* was discovered beneath one of the windows of the Marine Institute in Monte Carlo, Monaco. This alga, a native of Martin Bay, Brisbane, in Australia, was reared as an aquarium alga for tropical fish and was first reported in Europe (in Stuttgart Aquarium) in 1969. The alga, although certainly not genetically modified, had inevitably undergone some genetic mutation in the closed environment of the institute's aquaria. It would appear that an aquarium and its contents were

▲ Typically the coastal reefs are smothered in sea fans, algae, fish and invertebrates.

emptied out of the marine institute's windows and this alga, hitherto unknown in the Mediterranean, started to grow.

This act, originally accepted by the authorities, was soon dismissed as it was thought that the alga could not survive in the northern Mediterranean's colder water temperatures during the winter months. The director of the institute, Professor Doumenge, stated that this was purely a temporary phenomenon. However, as the alga started to spread, its arrival was then described by the French authorities as being some accidental import, either owing to climate change or even a Lessepsian visitor via the Suez Canal from the Red Sea.

▲ Sea Fern (*Halopteris scoporici*) dominates the shallow offshore reefs of the central Mediterranean.

It was soon discovered that not only were there no known species locally that would feed on this alga, but that it spread through fragmentation. Ships' anchors and chains, and fishermen's nets were inadvertently dragged through it, breaking off leaves and stems: and when the offending equipment was put back into the sea in another section of the coast, the broken pieces immediately started to grow. The Côte d'Azur was soon dotted with this rapidly spreading alga. Professor Alexandre Meinesz, a biologist from the University of Nice, was called in to assess the probable damage and subsequent spread of the alga. After 18 months of intensive research, Meinesz concluded that there was no way of stopping this ecological disaster. Six years later, Professor Meinesz stated that *Caulerpa taxifolia* was like a metastasising cancer whose spread could not be halted, and which could eradicate all Mediterranean life. He said: 'The water of the Mediterranean does not drop below 10°C; at this temperature, the alga is in stasis and can survive quite happily for over three months; at 15°C, the alga starts to grow quite rapidly; at 20°C, *Caulerpa* grows at over 3cm (1in)

a day'. Monaco, despite its commitment to marine conservation, has lost some 90 per cent of the seabed to *Caulerpa*.

In just five years, the initial 1m² (10ft 76in²) patch had already grown to one hectare. By 1990, it was discovered by divers off Cape Martin on the Italian border and this initial area now covers more than 3ha. The number of reported cases increased tenfold in the next year. In 1992, the alga was discovered in Majorca, Italy and Corsica, covering an approximate area of 420ha. By 1997, it was found in Croatia around Split, Messina in Sicily, and the island of Elba and now covers 4,700ha. At the same time 300ha of small patches of *Caulerpa taxifolia* were found at Sousse in Tunisia, and by 2001 this area had expanded to more than 13,000ha. Whenever new sites are found by divers or reported by yachtsmen or fishermen, the report is verified by the university and added to the mapping of 'Carpet Algae'.

Outbreaks of *Caulerpa taxifolia* have since been discovered off San Diego in California and in New Zealand. UNESCO have been informed of the catastrophic effect that the alga is having on the future of the Mediterranean, but as yet, the French authorities have still not come up with the cash or the personnel to try and extract the weed from the sea. One way of killing it is by spreading a large black plastic sheet over the affected area and then treating the alga with chlorine. This also kills off all other types of life, but the regeneration over time by Mediterranean species is preferable to the continual destruction caused by *Caulerpa*.

The alga is not only overtaking all local marine life; as a by-product of its growth, it creates a mud residue, further smothering this fragile ecosystem. A few small nudibranchs have been discovered that feed on this alga, but nowhere near the numbers needed to halt the spread. Small bream are also starting to eat the alga; however, the plant is so toxic that it will take several generations of fish before they are entirely immune. Then there is the problem of human health, should these toxic fish appear on the fish market.

However, early indications do show that the plant can be killed off by various methods and these trials have been successful in California and some areas of Spain. The sea slug genus *Elysia* is known to eat the alga exclusively and several types of copepods also thrive on the alga. It is hoped that with continued monitoring and the immediate response by all Mediterranean government agencies, this ecological disaster in the making may be averted.

THE HABITATS

Corals are not overly evident in the Mediterranean and, owing to the changing water temperatures during the year, there are no true coral reefs, made up of the stony hard corals more associated with tropical coral reefs. As such, the rocky 'reefs' are generally composed of ancient limestone blocks from some past age when corals were more prevalent in the Mediterranean.

These are now colonised by a living mantle of a mixture of small corals, sea fans, sponges, other cnidarians and various algae. Rocky coastlines and small islands have become encrusted with marine life and many species are found only in the Mediterranean. Much of the classification of the world's marine animals has come from the early observations made in the Mediterranean.

The habitats in this sea are as diverse as they are in any specialised environment. Although there are some very deep basins, we shall only concern ourselves with the veneer, which we are able to see on any visit to this ancient sea.

Rockpools

These areas, water trapped in pockets of rock after a receding tide, are miniature oases of life, often containing creatures that are able to withstand both a rapid change in

◄ Mgarr ix-Xini on the island of Gozo is perfect for night dives.

salinity (dilution after heavy rainfall, or concentration after evaporation) and differences in water temperature (at the height of the summer the water temperature in shallow pools can rise dramatically for the 6–8 hours before it gets covered by the advancing tide once more).

Sand

Typically, casts of marine worms are seen on sandy shores in most places. You may also see the tracks of marine snails and other creatures, and this region is home to flounders, skate and rays. Many burrowing creatures also occupy this habitat, such as starfish, urchins, anemones, eels and gurnard. The east coast of Spain, Syria, Lebanon and Israel, and much of North Africa, has some of the best beaches in the Mediterranean and their pure white sands are wonderful. Many islands have huge sand dunes formed over millennia as wind-blown sand piles up on the shore.

Mud

Deeper waters, and those affected by river run-off, build up a muddy seabed owing to the amount of detritus sedimentation, which gets washed into them during periodic winter storms and heavy rainfall. Burrowing animals such as various molluscs, sea pens, mud crabs, brittle stars, fireworks anemones, langoustine and burrowing anemones are typical of this habitat and are all excellent photographic subjects.

Rocky shores

These offer a firm substrate base onto which algae and other sedentary organisms can attach. The particular selection of marine life found will depend on the type of rock. Some soft rocks can be bored into whilst glaciated granitic rocks only support a few animals as there is nothing for them to get a firm grip on. Exposed locations will always be beaten badly by adverse weather conditions and this factor will also determine the types of marine life found. Encrusting algae, chitons, sea urchins, limpets and barnacles are perhaps the most common rocky shore inhabitants.

Rocky cliffs

Submarine cliffs are well known for their colourful fields of jewel anemones, under-hanging ledges covered in parazoanthids, Precious Coral (*Corallium rubum*), huge forests of sea fans, tube worms, sea squirts and other more sedentary creatures. But it is more the nature of the habitat which is appealing to divers, as there are often caves and caverns associated with these cliff faces which, when combined with a deep vertical wall, undoubtedly increase the enjoyment of the dive.

▲ Artificial reefs along the Cote d'Azur are havens for all manner of marine life.

Shipwrecks

We all know the attraction of such edifices, but these are also important habitats, quite often found in areas which would be otherwise rather poor in marine life. Wrecks provide an important holdfast for soft corals, sea fans, sponges, sea squirts and algae, and are also home to many different species of commercial fish. Many of the wrecks are, however, in deep water and quite often in areas of strong tidal currents. This limits diver intrusion, but also these wrecks would not have the same amount of marine-life colonisation without those limiting factors. Many ships (wrecks) are deliberately sunk as artificial reefs, to greatly enhance the scuba diver tourism dynamic.

◄ The Blue Hole in Gozo is calm, even when the outside waters are rough.

▲ Caverns are perfectly safe for finding different species of marine life.

Caves and caverns

Wherever the coastline is made up of a mixture of ancient limestone or sandstone and where geological upheaval has shifted the Earth's rock strata, caves and caverns are found. Some of the most spectacular are found in north-eastern Spain, the Balearics, Corsica and Sardinia, Croatia, Malta and Gozo. Huge vertical clefts cut through headlands and tunnels harbour marine species which might otherwise only be found in very deep water. Creatures which favour these dark conditions are sponges, slipper lobsters, small shrimps and Precious Corals.

Piers

The most obvious of the artificial structures which abut the seashore, the piers and breakwaters found around the Mediterranean are quite often the only shore access on many stretches of coastline, as development has irrevocably changed the original coastline in the pursuit of more

living space, commercial enterprise and tourism facilities. These structures have evolved over the centuries as boat traffic has altered in shape, size and frequency of visits. When all else fails and you want a reasonable dive in fair conditions, with a super-abundance of marine life, then the piers and breakwaters will provide this, usually including octopus, cuttlefish, anemones, sponges, mussels and numerous other invertebrates which live on or in the encrusting algae. You may have to gain special permission to dive piers and breakwaters in sensitive areas.

Posidonia meadows

Fields of Neptune Grass (*Posidonia oceanica*) are very typical of most Mediterranean habitats and are 'indicators' (indicating the presence) of clear, clean, oxygen-rich water. Fields of *Posidonia* are important nurseries for fish and shellfish and are home to many other associated species. These meadows are a vital and indispensable habitat.

Caulerpa

Fields of the newly introduced *Caulerpa taxifolia*, on the other hand, are seen as a plague along the southern coast of France, Monaco, Corsica and Majorca. This introduced species is spreading uncontrolled through the Mediterranean, smothering indigenous Neptune Grass and also beneficial algae. The presence of *Caulerpa taxifolia* means that all of the (local) marine life species and habitats may be under severe threat of extinction.

Tidal rapids

Common in a number of localities between various islands, tidal rapids are home to sponges, algae and other invertebrates which feed on the swiftly moving plankton passing on their way four times each day. Here the seabed is a mass of life and is considered a high-impact zone. Large populations of pelagic commercial fish and cetaceans also favour these areas.

REEF CONSERVATION AND THE TOURIST

Marine conservation is not new in the Mediterranean. Most of her bordering countries have specific marine parks, many of them more than 40 years old, providing important nursery areas for many different groups of fish and invertebrates. Commercial fishing is still concentrated at the perimeter of these marine parks, as the protected areas can only support so much marine life, allowing the 'overspill' to be fished. One zone in particular off Croatia is the 3,000km² (1,158 square miles) zone around the Jabuka/Pomo Pit area, which is a vital spawning ground for hake and langoustine. Protected since 2017, this zone is so successful that commercial fishermen are able to have thriving businesses around the perimeter of the zone.

Unfortunately, many important habitats and breeding areas have already been severely depleted and in some cases totally destroyed. Offshore islands have generally been spared from commercial activities, but the coastline bordering the Mediterranean has been under threat since people first inhabited those shores. Many offshore reefs and rocky shoals have been overfished. The catches of fish in general have been reduced to alarming levels and traditional fisheries have collapsed in many areas.

The breeding areas of the critically endangered Mediterranean Monk Seal (*Monachus monachus*) have been so depleted that there are now thought to be only around 700 animals left in the wild. They live mainly around the Greek Islands of the Adriatic, with an isolated group off Madeira.

The traditional tuna trapping or *mattanza* (from the Spanish word for 'slaughter') in Sicily and Sardinia still occurs each year as the Common Tuna (*Thunnus thynnus*) migrate from the Atlantic to their

▲ Old tuna net anchors lie abandoned in Tarifa, southern Spain.

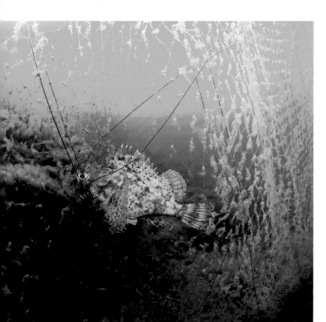

waste products. Grey mullet fisheries in Egypt have been successful, as these fish are vegetarian and also feed on detritus.

Mussel and oyster culture is responsible for the introduction of a further 60 species of algae and molluscs from the Japanese archipelago, all of which have been accidentally released into the wild from the seedling stocks imported from outside the Mediterranean. Many people depend on the sea for their livelihood, yet even these traditional ways of life are constantly under threat from the more commercial aspects of fishing for the mass market.

spawning grounds in the Mediterranean. These catches are now affected by Asian longliners, which catch huge quantities of tuna and other large pelagics without reference to any authority in the Mediterranean.

More insidious fishing methods have also affected the Mediterranean's fragile ecosystems. Aquaculture has grown alarmingly, as the ever-increasing population and tourists require feeding. Some fish farms require huge amounts of fish meal, which subsequently contaminates the surrounding seas with nutrient-rich

Thankfully, most countries now accept that a successful tourist industry relies on strict conservation policies and the protection of natural wetlands and fisheries. For this industry to succeed and prosper, tourists also have to be aware of the impact that they can make, especially on small areas, therefore education at all levels is vitally important. Membership of conservation agencies is always a clear step into the understanding and protection of the marine habitats.

ENDANGERED SPECIES

A number of marine species in the Mediterranean are threatened, principally due to human activity. Although pollution will often be considered to be one of the main causes for the loss of a species in a specific area, serious pollution is so localised that this is rarely the case, especially as more strict anti-pollution measures have also helped clean up much of the coastline. Other dangers may be much more direct. Sport spear fishing has

wreaked havoc on local inshore populations of more sedentary larger fish, such as groupers. Overfishing and the systematic plunder of fishery resources have resulted in massive reductions in Brown Meagre (*Sciaena umbra*), tuna, grouper and crayfish. Trophy catches such as the giant Noble Pen Shell (*Pinna nobilis*) or the collection of living skeletons (Precious Coral) for the jewellery trade have all had a detrimental effect on the ecosystem.

Local destruction and degradation of seagrass meadows through the construction of new ports and marinas has led to a far more insidious threat from increased tourism, where repeated use of ships' anchors, devastating use of fishing gear and increased (local) pollution have all led to the loss of many *Posidonia* meadows, as this seagrass is unable to reclaim lost areas. Couple this with the alarming rate at which *Caulerpa taxifolia* is covering the northern Mediterranean shores, and it can be argued that all native Mediterranean species are under threat.

Precious or Red Coral
Corallium rubrum

Divers still lose their lives each year as they search increasingly further into deeper water and caverns for this small, fragile coral. Used exclusively in the tourist jewellery market, the brilliant red branches of this exquisite coral are a delight to see underwater and are thankfully protected in a number of areas.

Noble Pen Shell
Pinna nobilis

The book *Guinness World Records* records the giant Noble Pen Shell as the second-largest shell in the world (after the Giant Clam). This species of mussel can exceed 1m (3ft) in height and is traditionally found in *Posidonia* seagrass meadows. Known to grow for more than 20 years, it is collected by trophy hunters and also used in the jewellery trade.

▼ Precious Red Coral has been used widely in the jewellery trade since Roman times.

▲ The Long-snout Seahorse is an iconic fish to be found from the Mediterranean to the British Isles.

Crowned or Long-spined Sea Urchin *Centrostephanus longispinus*

Sea urchins have long been regarded as a delicacy in the Mediterranean and the Long-spined Sea Urchin has been collected for generations. However, a virus has wiped out much of the Mediterranean population and, although it is still quite common in the eastern Mediterranean, it is considered quite rare elsewhere (however, it has been recorded as far west as Gibraltar).

Dusky Grouper *Epinephelus marginatus*

The local grouper population in the Mediterranean has been decimated by spear fishermen. Fairly sedentary in habit and often curious by nature, groupers have been an easy target for spear fishermen over many years. Malta and Gozo, Turkey, the Greek Islands, eastern Spain, Italy and southern France have all suffered from this. However, areas such as the Medas Islands in northern Spain, northern Sicily and the areas between southern Corsica and northern Sardinia are home to very healthy populations of large groupers due to the instigation of conservation zones many years ago.

Neptune Grass *Posidonia oceanica*

Posidonia is found off all Mediterranean coasts from depths of 0–40m (0–130ft). It is a flowering plant, not an alga. It is home to myriad marine creatures and an important fish hatchery, the plants actually anchoring the seabed. As it is very slow-growing, at only 3cm (1in) per year, it would take 3,000 years to replace the vast meadows of *Posidonia* which have been lost through the construction of coastal marinas. Replanting programmes are ongoing in many coastal countries.

Seahorses *Hippocampus* spp.

Seahorses are among many species affected by the loss of *Posidonia* meadows and coastal construction. The loss of the habitat is perhaps the largest problem facing these enigmatic and curious fish, but there are other factors which also threaten their existence, such as indiscriminate fishing methods and the Asian pharmaceutical interest, with more than 20 million seahorses sold throughout the world each year.

Mediterranean Monk Seal *Monachus monachus*

Near extinction, the Mediterranean Monk Seal has been hunted mercilessly by fishermen over the last 50 years as it was seen as the largest single threat to fishermen's livelihoods. It is restricted to a few small islands between Greece and Turkey, but early initiatives in marine conservation now appear to be working. However, with a restricted gene pool, the species' future in the Mediterranean is bleak. The seals are also found in Madeira and a few isolated pockets of Morocco.

MAKING CHOICES IN MARINE CONSERVATION

When booking a holiday, research the area first and use only diving schools which are involved with their local marine parks and conservation initiatives.

Contact the appropriate conservation agencies to see if there is specific information on the areas where you may want to dive.

Ask your tour operator if they have an environmental policy, and if they contribute to marine conservation societies.

Make sure that dive shops and operators explain their specific conservation policies before the dive or snorkel, as this will undoubtedly help your awareness and lessen your impact on the marine environment. Follow the example of other conservationists and use biodegradable shampoos, dispose of your litter appropriately, use fresh water sparingly and try and further the conservation message.

SOUVENIRS

Collection of marine souvenirs is prohibited in most areas, so respect all local and international laws.

The collection of empty beach shells may well seem innocent, but these shells may be used by small blennies as nesting sites, or by hermit crabs looking for a new home. Be careful in your selection.

All corals and turtle products are protected under CITES (the Convention on International Trade in Endangered Species)

and can only be bought and sold under licence.

Never buy marine curios, as they will probably be from another area of the world which is under even more threat and where the fishing methods used in the collection of the species may well be suspect.

▼ A diver explores an amazing variety of marine life including delicate sea fans.

SEA STINGERS OF THE MEDITERRANEAN

As far back as written records began, people have recorded the plight of their fellows following a sting from one or another of the denizens of the deep. Aristotle first accurately described the stinging properties of stingrays and jellyfish in 350 BC. The Greek poet Oppian wrote that the barb of a stingray could kill a tree and, on the same note, Pliny, describing the stingray, wrote in the *Historia Naturalis*: 'So venomous it is, that if it be struchen into the root of a tree, it killeth it: it is able to pierce a good cuirace or jacke of buffe, or such like, as if it were an arrow shot or a dart launched: but besides the force and power that it hath that way answerable to iron and steele, the wound that it maketh, it is therewith poisoned.'

Venom, in most cases, is used purely for defensive purposes. As a working underwater photographer specialising in marine life studies, I quickly learned that if the creature that I was approaching does not swim away, then it must have some other form of defence. These creatures include small coelenterates and jellyfish, fire worms with their tiny bristles, and various fish, anemones, sea urchins, starfish, molluscs and corals that possess pointed, modified tips of fins or other structures. In general terms, stinging mechanisms found around the mouth parts are for offensive reasons (primarily to disable prey), while stinging parts found along the back and tail are defensive in origin.

Different creatures have different types of toxins and potency. Most cause localised effects such as numbness, irritation or paralysis; others kill nerves and blood cells, attack muscles, or affect internal organs. Most have a cumulative effect and cause several problems at the same time. In certain circumstances, these toxins can even cause death in humans. Around 50 deaths each year are attributed to sea stings of some nature.

As already mentioned, one of the most common stingers are jellyfish. These come in a vast army of different sizes and potency. All are members of the same taxonomic group, which also includes corals and anemones. These creatures are primarily offensive stingers. Their stinging mechanism is in the form of a hooked barb fired by a hydraulic coiled spring. These barbs are called nematocysts and are held inside a trapdoor until they are released by touch or chemicals in the water. The barbs are hollow and filled with toxins which are released as soon as the stinger penetrates its victim. The primary aim is paralysis, before ingestion. Most jellyfish sting, but few are dangerous to humans.

When seasonal changes are in their favour you can encounter the Portuguese Man-of-War (*Physalia physalis*), a relative of jellyfish, in tropical waters. These are highly toxic and continued exposure to the stinging cells may require hospital treatment. Whenever the conditions are favourable for Luminescent Jellyfish (*Pelagia noctiluca*), there is always the chance of swimming beaches being

◀ The venomous sting of the Common Stingray was first described by Aristotle, Oppian and Pliny.

affected. Swimmers should wear protection such as a wet suit, rash vest or the new style of Lycra skin suit. There are local remedies available for stings, but acetic acid (vinegar) is as good as anything. In cases of severe stinging, medical attention will be required. Closely related to jellyfish are the anemones, hydroids and corals, all of which have a surprisingly large number of harmful representatives.

Most anemones will not do any harm to the much thicker skin on your fingers, but many can inflict quite painful 'burns' on softer

▲ Fire Worms (*Hermodice carunculata*) are well named as the tiny hairs on the body can produce a fiery hot stinging rash and should be avoided at all times.

parts, such as the inside of your arms or legs. Species such as the Berried Anemone (*Alicia mirabilis*) have warty tubercles all over the stem of the anemone; each 'berry' is armed with lethal nematocysts. The more common anemones use their barbs to hook and paralyse prey which swim inadvertently within their 'sticky' grasp.

Hydroids such as the Sea Nettle (*Pennaria disticha*) have harmless-looking, feather-like plumes which can inflict a rather nasty sting on the softer areas of your skin if you brush up against them. Even the most innocuous-looking sea creatures often have a hidden battery of stingers just waiting for something to rub against them. A few sponges have tiny calcium spicules which when rubbed against have a very similar effect to that of fibreglass rubbed against the softer parts of your skin. This can cause severe irritation, rashes and sores. Bearded Fire Worms (*Hermodice carunculata*), although quite cute in appearance, should never be handled. These attractive, small worms have clumps of white hairs along their sides which display bristles when touched. These bristles easily break off in the skin, causing a painful burning feeling and intense irritation. Although they are not deadly, stings will require treatment, principally with hot water and vinegar.

Perhaps the species we most associate with sea stings are members of the stonefish and scorpionfish families. There are now records of stonefish in the Mediterranean, along with several species of scorpionfish, all of which are armed with a mild, toxic venom in the modified hollow spines found at the tips of the dorsal fins. These are not considered dangerous but care, as always, should be taken to avoid the spines on the top of the dorsal fin. Inadvertent stinging can be helped by placing the affected area into very hot water. Common Lionfish are now prevalent in the eastern Mediterranean and they have venomous tips to their fins. (I have been stung by one and never want to repeat that experience!)

If you encounter a stingray at close quarters, you must never attempt to grab hold of the tail or sit or stand on its back, as the stinging mechanism is located in the tail. Any undue force on the creature may cause it to spring its tail forward in a reflex action, thus erecting the spine and causing serious damage. There are also species of electric ray found in the Mediterranean, which should be avoided. Weaverfish (*Trachinus* spp.) like to inhabit shallow coastal areas and lie partly buried in the sand. They have a venomous spine on top of the dorsal fin and, although not lethal, this sting can cause extreme discomfort when stepped on. The Stargazer (*Uranoscopus scaber*) has two venomous spines, one situated on each side behind the gill covers.

Not all stingers are large and obvious;

quite a large number of molluscs also have stinging mechanisms. Nudibranchs, for instance, eat stinging hydroids and anemones and have the ability to store the stinging nematocysts of their prey in their own tentacles. When attacked by predators, they are able to utilise the stored nematocysts in defence.

Sea urchins and starfish are the last and most obvious group of sea creatures that seem to lie in wait for unwary and clumsy humans. Thankfully, with good buoyancy control and using cameras with longer lenses, which allow them to photograph these spiny creatures from a distance, divers are now able to avoid most brushes with these animals. The spines of a number of sea urchins can be venomous. Even if not, they can puncture the skin – even through gloves – leaving painful wounds which can become septic. Rare now in the Mediterranean (owing to an epidemic which almost wiped out the entire population) is the Long-spined Sea Urchin (see page 158). This urchin should be avoided as the spines are quite brittle and easily broken off in the flesh. Made of calcium, the spines in the flesh should dissolve after a few

▲ The Common Lionfish is now spreading throughout the Mediterranean, having come from the Red Sea.

days. Deeply embedded spines may leave permanent scarring and patients may have to be treated for shock. Treatment includes rubbing the wound with the juice and pulp of the pawpaw, or even urine!

Most wounds to the unwary and uninitiated are caused by ignorance. As mentioned earlier, if the creature does not retreat from you, or if it exhibits bright colours or moves into a defensive posture, then you can be fairly certain that it has a defence mechanism which can harm you. For those unfortunate enough to encounter some of the pelagic stingers of the sea, such as tiny microscopic jellyfish, all the advice that you have will never prepare you for the agony. It is recommended at all times never to scuba dive in just a swimming costume. You should always wear either a full wet suit or Lycra 'skin suit' for overall protection. If you are cautious and careful with your buoyancy, you should be able to gain a new understanding and appreciation of these much-maligned 'stingers of the sea'.

DIVING AREAS

Gibraltar

Starting from the extreme west and working along the northern shores of the Mediterranean, the northern 'Pillar of Hercules', or Gibraltar as it is more commonly known, is at one of the world's natural crossroads. It is here that the incoming current from the Atlantic Ocean mingles with the denser waters of the Mediterranean Sea, creating a unique combination of flora and fauna. The 'Rock', as it is known, is home to several superb wrecks dating back to Napoleonic times, with the best of the wrecks being located off the breakwater and dating from the Second World War. These wrecks are home to a huge amount of marine life, including sea fans, sea cucumbers, schools of anthias, nudibranchs, octopus and cuttlefish.

Spain

Much of the southern Spanish coast is very similar in topography and species diversity to Gibraltar. As you travel north-east towards the French coast, there are a number of marine protected areas, with good diving and excellent marine life to be found at Al Muñequa and Fuengirola. The most famous of all the protected areas is the Medas Islands, off the coast near the resort of Estartit. These small rocky islands have been protected since the early 1980s and have absolutely huge concentrations of fish, including grouper, bass, bream, sardines and mullet. The mainland coast is carved with hundreds of gullies, caves and caverns, many of which travel several hundred metres underground and are home to slipper lobsters, Unicorn Shrimps (*Plesionica narval*), colourful sponges, Spiny Lobsters (*Palinurus elephas*) and sea hares.

Balearics

The Spanish islands of Formentera, Ibiza, Majorca and Menorca all have excellent diving, with many fine diving schools. The small island of Cabrera is undeveloped and is now a national park. The entire area has a similar feel to Malta and Gozo, with deep caverns, huge burrowing anemones, plenty of algal turf being grazed by wrasse, but few big fish. The best diving in Majorca is along the mountainous west coast between the ports of Polense and Andratx and there are huge schools of Mediterranean

▲ Gibraltar is at the confluence of the Atlantic and Mediterranean.

▲ The Medas Islands are one of the oldest marine reserves in the Mediterranean.

Barracuda (*Sphyraena sphyraena*) at the tip of Dragonara Island. Pont D'en Gil cavern in Menorca is popular with divers due to the ancient stalactite and stalagmite deposits underwater – testimony that this cave was once on dry land, before the Mediterranean was flooded.

France

There are a number of marine parks all along the southern French coast, starting with the nature reserve of Cerbère-Banyuls just north of the Spanish border. Very similar geologically to Estartit, with many sea caves, this reserve is supported by the Arago Oceanographical Laboratory, located at

▲ Located midway between Cap d'Antibes and Iles de Lérins is an isolated series of rocky pinnacles, topped by an automated lighthouse called *La Fourmigue*.

Banyuls-sur-Mer to the north. The reserve is noted for its Dead Men's Fingers (*Alcyonium palmatum*) and Precious Corals. The island of Porquerolles, near Toulon, is administered by the French National Trust and has many interesting wrecks nearby. The Côte d'Azur has excellent diving around the offshore islands near Cannes, Cape Juan and Cape Antibes with some superb walls and interesting topography. The bay at Villefranche-sur-Mer has numerous small caverns where small Precious Corals can be found.

Monaco

Known as the home of the famous Oceanographical Institute supported by Jacques Yves Cousteau, Monaco still promotes itself as a leader in world conservation policies. However, it has lost 75 per cent of its natural coastline due to development. Monaco is now seen as being directly responsible for the introduction of *Caulerpa taxifolia*, which is endangering the entire Mediterranean. Thankfully all of its coastline is protected now, and strict conservation policies are enforced.

Corsica

Cap de Scandola on the north-west coast of Corsica is set in the heart of a marine park that is dominated by huge, tortured, red volcanic rocks that plunge into the sea. Difficult to get to, the diving in the marine park is dominated by fields of Red (*Paramuricea clavata*) and Yellow (*Eunicella cavolinii*) Sea Fans. Off the south-east coast can be found the Lavezzi Islands Nature Reserve which has been protected since 1970. The islands, of granite formation, are renowned for their friendly populations of large grouper.

Sardinia

Sardinia is under the protectorate of Italy and on the other side of the Strait of Bonifacio can be found the twin island nature park of Maddalena Island, also favoured by fish-watchers. South of this small archipelago can be found the Gulf of Aranci at Capon Figari, where large concentrations of the Noble Pen Shell can be found, some of which are almost 1m (3ft 3in) tall. Off the north-east coast can be found the spectacular cave and cavern formations of Nereo Cave at Capo Caccia. There are also a number of scenic wrecks to be explored along the east coast of the island.

Italy

Near Santa Margharitta at San Fruttuoso can be found the underwater sculpture of the *Christ of the Abyss*, placed by Guido Galletti in 1954. A copy of this statue resides in the Pennekamp State Park in Florida. Commemorating the life of Cressi Sub founder Egidio Cressy, the statue is in the centre of a fine marine park which is home to thousands of damselfish. The island of Giannutri midway down the west coast of Italy has a couple of fine wrecks and nice sheltered bays. The *Nasim II* and the *Anna Bianca* are both in deep water, but are surrounded by thousands of anthias. The next group of islands to the south includes Ponza, where the wreck of the *LST349 MK. III* lies. Sunk in 1943, the ship is an eerie reminder of the last World War. On nearby Ventotene Island can be found the wreck of the *Santa Lucia*, which also dates from 1943, but is a much better wreck for marine life.

Sicily

Bordering two major sea areas, Sicily forms part of the ancient land bridge which once split the Mediterranean. Not widely known for its conservation policies, Sicily has made more obvious steps in recent years with the tiny island of Ustica, just 58km (36 miles) off the coast of Sicily, facing Palermo. The tip of a subterranean volcano, the island is known for the huge amount of Sea Rose (*Peyssonnelia squamaria*) which can be found everywhere. Large colonies of Red Sea Fan are also evident. On the east coast, near Catania, is the marine reserve of Aci Trezza. Founded in 1992, the reserve is studied by the University of Catania, and more than 300 species of algae have been recorded within the park's boundary. The islands are said to be the historical 'Cyclops Islands'. Legend has it that the island is actually the colossal stones flung by the giant Polyphemus at Ulysses' boat.

Malta, Gozo and Comino

These islands are located between Sicily and Libya – their position is responsible for their huge diversity of marine life. The country's

▼ Xlendi Bay in Gozo is a superb night diving area where Filefish and Golden Eels can be found.

first marine reserve was declared in 2011 around Cirkewwa, close to the ferry terminal between Malta and Gozo. The islands have superb clear water, fantastic caves and caverns and many historical wrecks including aircraft from the Second World War. Numerous alien species from the Caribbean are being found more recently, including filefish. Common Lionfish are now also in these waters.

Croatia

On the border of the north-eastern Adriatic, the islands off the coast of Croatia have been a sailor's delight for years. Now divers are exploring beneath the

▲ The Mediterranean Monk Seal is endangered but small colonies are found around the Greek Islands.

waves, particularly off the island of Korcula, and although the area has been overfished for many years, the invertebrate populations are high, with many different hermit crabs and nudibranchs. Large sponges, Feather Starfish (*Antedon mediterranea*) and Red Sea Fans are symbolic of the diving off Korcula. Great White Sharks (*Carcharodon carcharias*) have been spotted near Istria, following the tuna migrations. There are several sites which have ancient *amphorae* (clay storage jars) as well as wrecks from the Second World War, including a fairly intact German submarine.

Greece

The Greek Islands are said to be the cradle of civilisation, and there are countless numbers of historical monuments and ancient wrecks to testify to this fact. Sport diving was once restricted to a few select sites of little interest. However, over the last few years, more areas have become available for exploration and now divers are enjoying the wealth of marine life and can also visit historic sites and underwater artefacts. The endangered Mediterranean Monk Seal is also found in this region.

Crete

Separated from Turkey by the Sea of Crete, this ancient Greek island has some interesting diving around its rocky shores. Best for diving is the western shore, which is less touristy. Nearby Chania has a large area of ancient *amphorae*, now covered in an algal turf. There are large schools of bream and wrasse as well as interesting caves and caverns.

Cyprus

Diving is very popular in Cyprus, particularly around the Paphos region, with a number of diving schools having excellent reputations. The most famous wreck in Cyprus waters is that of the *Zenobia*, which sank in 1980. This 172m (570ft) wreck is a must for all divers visiting the island. The Akamas Peninsula is now earmarked as a marine park and the Akrotiri Fish Reserve near Limassol is very popular.

Turkey

South-eastern Turkey has some superb diving, especially around the popular holiday resorts of Fethiye and Marmaris/

Icmeler. Although it is not noted for its fish life, the huge sponges, octopus and other invertebrates more than compensate.
The region is also dotted with ancient shipwrecks, sea caves and caverns, perfect for exploration. Turkish divers also know of a certain area where you can encounter Dusky Sharks (Carcharinus obscurus). More Red Sea species are being recorded from these waters than ever before.

Syria

There is little knowledge of diving off the coast of Syria, but from accounts by Turkish divers the topography is very similar, with an interesting mix of fish coming up from the Red Sea. Diving here would have to be done by special permission and then only by boat.

Lebanon

Lebanon has a number of good wrecks, mainly dating from the civil war that shook the country in 1975. The country is also known to be home to small numbers of Ragged-toothed Sharks (Carcharius taurus). Please note that at the time of writing, foreign travel is restricted due to political instability.

Israel

There is some inshore diving to be found on the Israeli Mediterranean coast, but it is fairly restricted and generally on some not too ancient wrecks. These are all well colonised now by different algae and small anemones. Fish life is interesting, as there are an increasing number of migrants from the Red Sea to be found. The three most interesting dive areas are: a Second World War Italian Navy submarine; an area near a power plant at Hadera which has a large number of sharks in the winter months that are attracted by the warm water discharge; and the archaeological site at Caesarea.

Egypt

The Egyptian coast was not known for its diving until some archaeology researchers found ancient remains off the harbour wall in Alexandria. These ancient obelisks, sphinx and statues are surrounded by damselfish

and bream with a mix of fish and algae from the Red Sea. The area is quite unusual and has opened up for scuba divers to explore.

Libya

The recent political troubles mean that the future of Libya is uncertain at present. However, prior to this the country was newly opened for tourists. Although diving was still in its infancy, the early reports stated that the offshore reefs and wrecks were pristine. Many of the larger inshore fish have gone, but there is a large invertebrate population.

Tunisia

The islands of Zembra and Zembretta were declared a national park in 1977 and are included in UNESCO's list of Reserves of the Biosphere. The rare Giant Limpet (Patella ferruginea) is found here as well as local groups of dolphin. The National Institute of Marine Sciences and Technologies, which was founded in 1927, is involved in continuous research on the islands. Founded in 1980, the second reserve includes the islands of Galite and Galiton off the northern Tunisian coast and consists of one main island and five smaller ones. Of granite origin, unique in Tunisia, the islands are renowned for their Posidonia meadows, crayfish, large grouper and Precious Coral. They were once home to the Mediterranean Monk Seal, and it is hoped that some may colonise the islands once more.

Algeria

Little is known about the Algerian coastline, but there are wrecks from various political conflicts, as well as sunken cities and good marine life to be explored.

Morocco

Owing to constant political changes in the country, there is little scuba diving done. However, several operators from Gibraltar and southern Spain do occasionally cross the straits to dive some of the offshore reefs and islands. These are known for large grouper, lots of sea fans and a wide mix of Atlantic marine life.

Protect the ocean

Ten ways a diver can protect the aquatic realm (produced by the PADI Aware Foundation):

- Dive carefully in fragile aquatic ecosystems, such as coral reefs.
- Be aware of your body and equipment placement when diving.
- Keep your diving skills sharp with continuing education.
- Consider your impact on aquatic life through your interactions.
- Understand and respect underwater life.
- Resist the urge to collect souvenirs.
- If you hunt and/or gather game, obey all fish and game laws.
- Report environmental disturbances, or destruction of your dive sites.
- Be a role model for other divers in diving and non-diving interaction with the environment.
- Get involved in local environmental activities and issues.

▼ A typical zonation of an offshore reef with sea grass at the top gradually being replaced with algae and sponges.

KEY TO SYMBOLS

This key describes the symbols that appear at the head of each species description. The symbols give a quick guide to the habit, diet and habitat of each species.

HABIT

 single

 pairs

groups

schools

DIET

 algae

 plankton

 invertebrates

 mixed corals

 fish

 mammals

 turtles

HABITAT

 caves

 seagrass

 sand

 reefs

 pelagic

 surface waters

CLASSIFICATION AND NOMENCLATURE

You will note that in this guidebook to the Mediterranean, both English and scientific names are provided for many species of marine life. The scientific name of any particular animal is very important. When diving in various parts of the world, or even in the same region, you may come across several different names for the same organism. This can be confusing. Scientists prefer that, when identifying or describing a particular animal or plant, you use its scientific or specific name.

The correct naming of a species is also very important for your own log book records and is essential to scientists and marine biologists studying flora and fauna, now and in the future. The modern binomial system of nomenclature (a two-part name, the first for genus and second for species) was developed by Carl Linnaeus and dates from publication of his *Systema Naturae* in 1758 and subsequent years.

Scientific names are written in italics, or even underlined in some texts. The first (genus) name always has a capital letter and is followed by the specific or trivial name in lower case, for example *Posidonia oceanica* (Neptune Grass). Once you get into the habit of using scientific names, you will soon discover how easy it is, and how useful it is to describe species in a universal language used by enthusiasts.

Almost all of the fish and invertebrates included in this guide have been described to species level. In species that differ significantly by sex and/or age, photographs of male and female, and juvenile and adult, are provided. However, some of the more difficult-to-identify organisms are described only to genus level.

HISTORY OF BIOLOGICAL RESEARCH

Pliny the Elder (*Gaius Plinius Secundus*), a Roman author, naturalist and natural philosopher who lived during the first century AD, and died aged 56 during the eruption of Vesuvius in AD 79, was one of the first writers to begin to catalogue Mediterranean life. The volumes in his *Naturalis Historia* include an encyclopaedia of animal and marine life, particularly from his native Mediterranean.

Surprisingly, there was little scientific study undertaken in the Mediterranean until the twentieth century. Pliny's observations were almost two millennia ago, and most of those were done from the surface and by dredging nets over the seabed or by hook and line. In the seventeenth century, the Italian nobleman Luigi Masili had speculated about counter-currents, but it wasn't until a two-year-long Danish expedition in 1761, which covered as much of the Mediterranean as possible, that the pelagic life associated with those counter-currents was properly studied. Only since the 1950s have any systematic exploration and observations been undertaken, under the auspices of NATO and various departments of the EU, including the work of Jacques Yves Cousteau from his research ship *Calypso*.

▲ Professor Meinesz from the University of Nice has done extensive research on Mediterranean algae.

EXTERNAL FEATURES OF FISHES

This diagram illustrates the main structures of a fish referred to in the species descriptions.

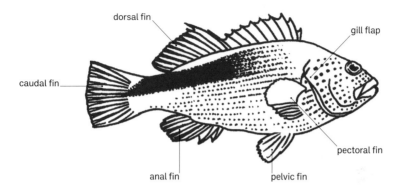

IDENTIFICATION GROUPS AND PICTORIAL GUIDE TO FAMILIES

Colour varies greatly between fish species, therefore it would seem an ideal means of identification. However, even within species, colour varies according to sex, age, region, season and surroundings. For this reason, body shape is a much more reliable means of identification. The following outlines represent most types of fish likely to be encountered in the Mediterranean. Those sharing similar characteristics are grouped together for initial identification.

Rays and Sharks

Carcharhind sharks (*page 36*)

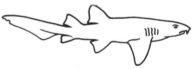

Lesser Spotted Dogfish and Nurse Hound (*page 40*)

Stingrays (*page 42*)

Electric Ray (*page 42*)

Regular-shaped fish swimming close to or above reef

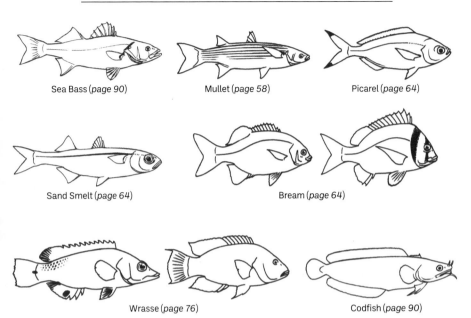

Sea Bass (*page 90*)

Mullet (*page 58*)

Picarel (*page 64*)

Sand Smelt (*page 64*)

Bream (*page 64*)

Wrasse (*page 76*)

Codfish (*page 90*)

Colourful regular-shaped fish close to reef

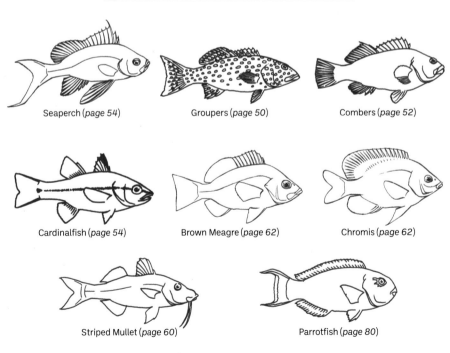

Seaperch (*page 54*)

Groupers (*page 50*)

Combers (*page 52*)

Cardinalfish (*page 54*)

Brown Meagre (*page 62*)

Chromis (*page 62*)

Striped Mullet (*page 60*)

Parrotfish (*page 80*)

Elongate fish close to or on reef surface

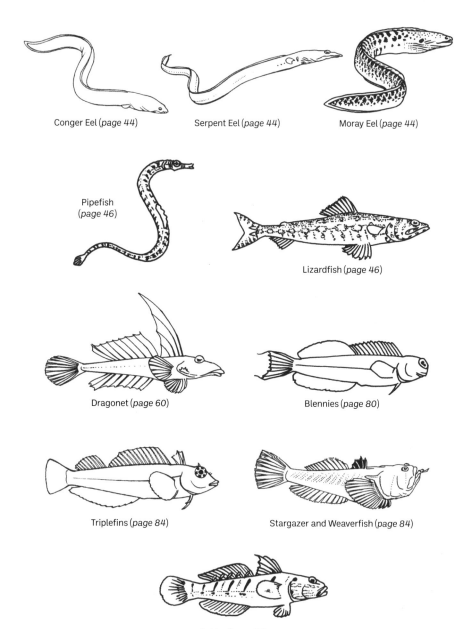

Conger Eel (*page 44*)

Serpent Eel (*page 44*)

Moray Eel (*page 44*)

Pipefish (*page 46*)

Lizardfish (*page 46*)

Dragonet (*page 60*)

Blennies (*page 80*)

Triplefins (*page 84*)

Stargazer and Weaverfish (*page 84*)

Gobies (*page 86*)

Silvery, torpedo-shaped fish

Blue Runner (*page 56*)

Jacks and Amberjacks (*page 56*)

Barracuda (*page 58*)

Irregularly shaped fish

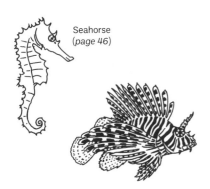

Seahorse
(*page 46*)

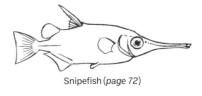

Snipefish (*page 72*)

Lionfish (*page 48*)

Anglerfish (*page 48*)

Rockfish and Scorpionfish (*page 50*)

Gurnard (*page 60*)

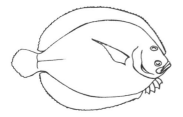

Flounder and Turbot (*page 92*)

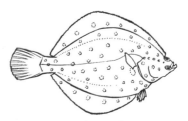

Plaice (*page 92*)

FISH

CARTILAGINOUS FISH (SHARKS and RAYS)

Sharks

THERE ARE TWO subdivisions of fish: the elasmobranchs, which are cartilaginous and include all the sharks and rays; and the teleosts, which comprise all bony fishes. Sharks are a highly specialised and ancient group, present in the Mediterranean. You are most likely to encounter the catsharks, which are not harmful to humans, but there are also Great White Sharks to be found. All sharks eat living animals, but many also eat decaying matter. Most are active hunters at night and all are shy. They are able to detect low-level electrical currents and can find sleeping fish at night. With highly developed sensory ampullae, they are able to detect low-frequency vibrations from great distances, as well as the scent of injured or dying animals.

1 GREAT WHITE SHARK
Carcharodon carcharias

The largest apex predator in our seas, the Great White Shark has been known to inhabit the Mediterranean for many years and has been considered responsible for a number of human disappearances and deaths. One attacked a small boat in the Adriatic in 1977 and the species really hit the headlines when Alfredo Cutajar landed a specimen 7m (23ft) long in the same year. Its main prey are the migrating tuna, and it is often caught during the traditional herding up or *mattanza* of these huge fish.

there is enough food in an area. It has been fairly well established in the Mediterranean for quite a number of years now, but is rarely found on shallow coastal dives. It is rather uniform grey with a white belly, and has dark stripes down its body which fade with age. Its blunt nose and forward-facing eyes afford clear vision and accurate judgement of distance. Growing to around 5m (17ft), it eats anything!

3 BASKING SHARK
Cetorhinus maximus

The second-largest fish in the sea (after the Whale Shark), this is a plankton eater and is very wide-ranging throughout the Atlantic and Mediterranean. It is easy to spot when feeding on the surface due to its tall dorsal fin and anal fins which also break the surface. Growing to around 9m (30ft), it uses its huge mouth to scoop up the plankton.

2 TIGER SHARK
Galeocerdo cuvier

The Tiger Shark is usually a solitary hunter, although small groups will come together if

1

2

3

1 BLACKTIP SHARK
Carcharhinus limbatus

This relatively small shark grows to only around 1.8m (6ft) and is more common in the southern Red Sea where it enjoys roaming along the edge of the deep reefs and vertical walls. It is brownish grey and has a white belly and distinctive black tips to all of its fins and tail. With the widening of the Suez Canal, it is well on its way to becoming much more established in the Mediterranean, and is found regularly in the eastern region.

2 SILKY SHARK
Carcharhinus falciformis

This is a pelagic shark, only growing to around 3m (10ft) and usually encountered around the offshore reefs and small islands. Often found in small hunting packs, it is quite curious and may come close by.

3 SAND TIGER SHARK
Carcharias taurus

The Sand Tiger Shark or Ragged-tooth Shark, which grows up to 3.30m (11ft) long, prefers deeper water where it feeds on a variety of bony fish and crustaceans. Although very fearsome-looking, it is not known to attack people. The species is now collected commercially for the aquarium trade as it can settle in large enclosed environments, such as sealife centres or public aquariums. However, there are problems associated with transport and ongoing health, and a number of aquariums have reported fatalities; conservation bodies are opposed to this trade. Regularly seen in the eastern Mediterranean during late summer, particularly off the coast of Lebanon, Syria and southern Turkey, it can occur quite close inshore near vertical cliffs and reefs. Denser than water, it is known to swallow air at the surface to help its buoyancy.

4 GREAT HAMMERHEAD SHARK
Sphyrna mokarran

Another pelagic shark, often found in small groups. It mainly feeds on the seabed, picking at detritus, dead fish and rays – its main prey. It pins down a ray with its wide cephalofoil or 'hammer', and eats it from above. Not known to be dangerous to people, it is often caught as bycatch.

5 COMMON THRESHER SHARK
Alopias vulpinus

Quite rare in the western Mediterranean, this shark is more common around some of the islands off Tunisia and further north to Turkey and Greece. A Lessepsian migrant from the Red Sea, it is a dark blue-grey above with a white underside. It has large dorsal fins and a very obvious elongated upper lobe to its tail, and can grow to around 5–6m (17–20ft).

6 SMOOTH HOUND
Mustelus mustelus

Growing to around 1.60m (5ft 4in), this small, smooth-skinned shark has large eyes and a long and slender body with a sharply pointed snout and five gill slits. It has two large, almost equal-sized, dorsal fins and, unlike the Spurdog, has an anal fin. Its main prey species are bony fish, squid and octopus, and small crustaceans. The species is mainly bottom-living and tends to stay in deep water during the day, rising into shallow waters around the coastline at night to feed.

1 SPURDOG
Squalus acanthias

Very similar in shape and appearance to the Smooth Hound, this small shark is distinguished by an obvious spur or spike directly in front of the dorsal fin. The male reaches over 1m (3ft 3in) in length, whilst the female is much smaller at 75cm (2ft 6in). Fairly slow-moving, it migrates from deeper waters into shallow estuaries during the summer months to have its young.

2 BLUE SHARK
Prionace glauca

This shark grows to 3m (10ft), has a slender body, wing-like pectoral fins, a long snout and large eyes, and is blue in colour. At night the colour appears more greenish. It feeds on small fish and squid, rounding up the schools and attacking fiercely. Under threat from human activity, it is caught by drift net and 'sport' anglers. This is a pelagic shark, often working in small groups as it hunts inshore shoals of fish. It is the widest-ranging of all cartilaginous fish, being found in all tropical and temperate waters. Tagged individuals from the Canary Islands have been found in Cuba and individuals from the United States have been captured in Gibraltar, indicating that they use the equatorial currents in their migration patterns.

3 LESSER-SPOTTED DOGFISH
Scyliorhinus canicula

Although this is referred to as a 'dogfish', it is actually in the catshark family and is perhaps the most common of all shark species encountered in the region. Rarely over 1m (3ft 3in) in length, its body is rough to the touch and is covered in dark spots and blotches. It is common in shallow waters and lays its eggs from May right through to September, attaching its 'mermaid's purse' egg case by four long flexible tendrils to various sea fans. Depending on the location and amount of plankton in the water, these egg cases gradually become covered in marine life, further protecting the embryo.

4 NURSE HOUND
Scyliorhinus stellaris

Very similar to the Lesser-spotted Dogfish, this is a much larger species, growing to almost 2m (6ft 6in). A purely nocturnal feeder, it preys on octopus, molluscs, crustaceans and demersal fish. It is collected for the aquarium trade as it is sedentary by nature and requires little work to keep it happy!

5 ANGEL SHARK
Squatina squatina

Growing to around 1.80m (6ft), this is a nocturnal bottom-dweller, more like a large ray than a shark. Roughly diamond-shaped, it has large, fleshy pectoral fins and a rounded snout with a speckled body. It is generally buried in sand and ambushes fish in this way, lying in wait until they pass close to the large mouth. It opens its mouth wide whilst lunging upwards, drawing the hapless prey into its mouth. It is quite harmless to humans and generally fearless of divers.

6 COMMON GUITARFISH
Rhinobatus rhinobatus

More of a ray than a shark, this is a common bottom-dwelling predator, feeding on benthic invertebrates and fishes. Growing to around 120cm (4ft), it has one or two litters each year with around 8–10 embryos. Commercially caught off the west coast of Africa, it is usually only caught as bycatch in the Mediterranean.

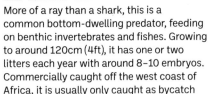

Rays

RAYS ARE INSTANTLY distinguishable as their pectoral fins are joined onto the head and form a broad skirt around the body. They have a cartilaginous skeleton like the sharks, but have no dorsal fins and the mouth is on the underside.

1 COMMON EAGLE RAY
Myliobatus aquila

This species is common in all areas of the Mediterranean, and the north-eastern Atlantic from the Azores, Madeira and African west coast to Namibia. Often coming fairly close to shore, it is encountered on many dives. It has an obvious snout with which it digs up the seabed in search of molluscs and crustaceans. It can grow to around 180cm (6ft) across the wingtips.

2 HONEYCOMB STINGRAY
Himantura uarnak

A Lessepsian migrant, this large ray is steadily spreading through the Mediterranean. It can grow to 1.5m (5ft) in diameter. The distinctive blending of the spots over its back into a honeycomb pattern makes it easily recognisable. It has an extremely long tail, up to twice the body length. The pattern gradually fades as the animal matures.

3 THORNBACK RAY
Raja clavata

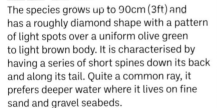

The species grows up to 90cm (3ft) and has a roughly diamond shape with a pattern of light spots over a uniform olive green to light brown body. It is characterised by having a series of short spines down its back and along its tail. Quite a common ray, it prefers deeper water where it lives on fine sand and gravel seabeds.

4 FOUR-EYED RAY
Raja miraletus

Another migrant from the Atlantic, this easily recognisable ray grows to 60cm (2ft) and has two or four distinctive dark spots on its back. Fairly common in all sandy areas, it is commercially caught and is commonly seen in fish markets. It is often seen on early evening and night dives, hunting in pairs.

5 COMMON STINGRAY
Dasyatis pastinaca

Widely distributed but quite rare, the Common Stingray usually grows to 60cm (2ft), although huge specimens have been recorded off Madeira and the Canary Islands. It is found locally in most areas of the Mediterranean and prefers a sandy bottom, quite often near the base of cliffs. Known from Ancient Greek times, it has one or more barbed spines at the base of its tail, which can be whipped forward and plunged into a predator in a purely defensive posture.

6 ROUND STINGRAY
Taeniura grabata

Very similar to the Common Stingray, this ray can grow to around 120cm (4ft) across and, as the name would suggest, is much more circular in shape. Found throughout the Mediterranean, its location would indicate that this species is also a Lessepsian migrant.

7 ELECTRIC RAY
Torpedo marmorata

This fish is easily recognised by its round body and thick fleshy tail. It is important to recognise the species, as its electric shock can stun a person. Light brown with a mottled or marbled appearance, its eyes are set far forward on the head with a pair of lobed spiracles directly behind. It grows to 60cm (2ft) long. When hunting, it folds its wings around prey and stuns it with an electric shock, recorded at 220 volts and 8 amps. This species is solitary and nocturnal.

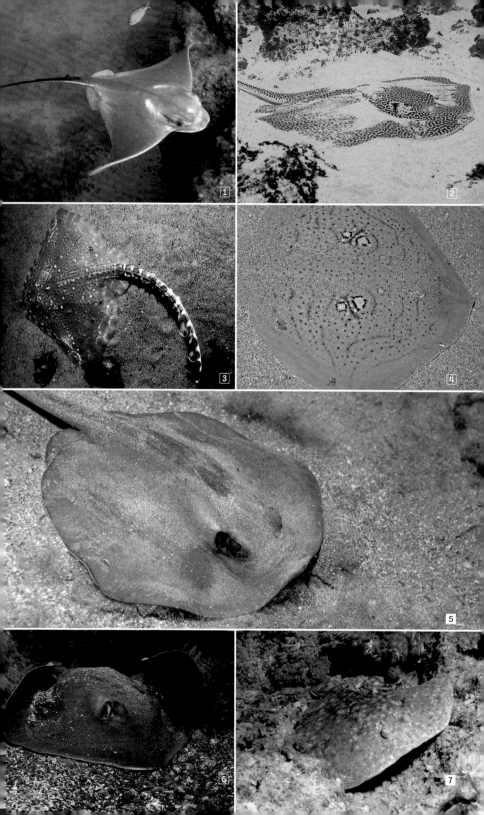

TELEOSTS or BONY FISH

MORE THAN 90 PER CENT of all other fish are bony fish, with a skeleton of bone rather than cartilage, and a large bony plate that protects the gills. Most also have distinct rays or spines on their fins. Ray-finned fishes form the sub-group Actinoptergii, while the lobe-finned fishes constitute Sarcopterygii.

Conger Eels
Family Congridae

1 CONGER EEL
Conger conger

The Conger Eel is greyish blue, cylindrical and can grow over 2m (6ft 8in) in length. It has a scaleless, snake-like body with a single long dorsal fin that merges into the tail and anal fins. It has large strong jaws and enjoys a diet of crustaceans and small fish. It is common throughout the entire region and is an inquisitive fish, tending to live in caverns or rocky crevices. It is also found regularly on shipwrecks, where there is always an abundance of long, fish-shaped pipes to hide in. Little is known of its breeding habits, but it is known to migrate into deep-water spawning grounds where each female lays several thousand eggs. It is similar in ecology to the Common Eel *Anguilla anguilla*, with young taking two to three years to return to the inshore habitat.

2 BANDTOOTH or GOLDEN CONGER EEL
Ariosoma baleriacum

Found throughout the Mediterranean and north-west Africa, this eel can grow to around 50cm (1ft 6in). It is an active feeder at night, with large, close-set eyes. It is commonly found on night dives. The body is greyish with a golden sheen.

Snake Eels
Family Ophichthidae

3 LONGJAW SNAKE EEL or SERPENT EEL
Ophisunus serpens

Ovoid in shape, this species grows to around 45cm (1ft 6in) long and has a light tan body with numerous brown spots at the front of the head. The eyes are close together at the front of the snout, which is quite long with a jaw filled with sharp teeth. This quite rare fish hides under the sand during day and pokes its head out at night-time, when it lies in wait for passing prey such as small shrimps and fish.

Moray Eels
Family Muraeidae

4 MORAY EEL
Muraena helena

The Moray Eel can grow over 1.5m (5ft) in the Mediterranean, and is a favourite fish for divers to find. It has quite a small head, brown in colour, and tan spots, which develop into broader golden markings on the flanks. It has a long dorsal and anal fin, which merge into the rounded tail, and no pectoral fins. The outer edge of the fin is spotted with either white or gold. Occurring on shallow reefs and wrecks all round the coast, this species tends to come out at night to feed and will hide in a protective hole during the day. It usually has a few symbiotic shrimps also living in the same hole, which keep the fish clean of parasites.

Catfish

① STRIPED EEL CATFISH
Plotosus lineatus

Eel-like in shape, this brown catfish has very obvious lateral cream-coloured stripes. The head bears the barbels typical of catfish and it has venomous spines on the first dorsal fin and pectoral fins. This Lessepsian migrant is now common in the eastern Mediterranean and can often be seen in large groups.

Pipefish and Seahorses

Family
Sygnathidae

② LONG-SNOUTED SEAHORSE
Hippocampus guttulatus

In Roman mythology, Neptune's chariot was drawn by beautiful white seahorses. The genus name *Hippocampus* is from the Greek *hippos* meaning 'horse' and *kampe* meaning 'worm' or 'caterpillar'. This curious fish has a horse-like head set at right angles to the body, the trunk of which is generally short and fat, tapering to a long prehensile tail. It grows to a maximum of 15cm (6in) and is usually found with its life-long mating partner. Seahorses feed on small shrimp and other tiny planktonic crustacea, 'hoovering' them into the small mouth at the tip of the long snout. They enjoy algae-covered rocks and seagrass meadows in well-illuminated and aerated water. They move infrequently, making them difficult to spot, and the bits of algae on their long fringed 'mane' only enhance their already excellent camouflage.

③ SHORT-SNOUTED SEAHORSE
Hippocampus hippocampus

Commonly found throughout the Mediterranean, Black Sea, the southern British Isles and north-west Africa, this species grows to around 12cm (4in) and lives around seagrass beds. It has no pelvic fins and is usually a fairly uniform colour, varying from near black to a very pale beige or purple. As with all seahorses, the male broods the young in a specialised pouch on its abdomen.

④ PIPEFISH
Sygnathus acus

Pipefish are closely related to seahorses and, as the name implies, are long and thin with tube-like bodies over 45cm (1ft 6in) long. This species' snout is more than half of the head length and there is a distinct lump on the head behind the eyes. The body is ridged with obvious scaly plates. The dorsal fin is set near the rear of the body and it has no anal or pelvic fins. Seen fairly regularly throughout the region, it is difficult to spot as it lies along algae fronds, rendering it almost invisible.

⑤ MEDITERRANEAN BROAD-NOSED PIPEFISH
Sygnathus typhle rondeleti

This pipefish blends perfectly well with its seagrass habitat, due to its coloration and broader snout. It often rests in an upright position, or lies amidst leaf debris. Growing to over 30cm (12in), it can be quite active in short bursts, but its camouflage renders it virtually invisible very quickly.

Lizardfish

Family
Synodontidae

⑥ ATLANTIC LIZARDFISH
Synodus saurus

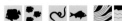

The Atlantic Lizardfish has a cylindrical body with a wide, slightly upturned head and a wide mouth absolutely jammed full of teeth. It has two dorsal fins, one very large, which it folds back against the body, the other tiny. It grows to 40cm (1ft 4in) in length. Its light tan and mottled brown appearance provides camouflage when it dives for cover into the sand. It is common in shallow sandy areas where it hides under the sand. Quite often, all that you will see are the two eyes and the upturned mouth.

Anglerfish

Family
Lophidae

1 ANGLERFISH

Lophius piscatorius

The Anglerfish or Monkfish is unmistakable with its large flattened body and incredibly ugly visage (unless you are another anglerfish!) Growing to over 2m (6ft 6in), it has a huge head with a frill of fleshy appendages around the lower jaw. The mouth is filled with sharp inward-pointing teeth and, when fully opened, is virtually circular. Behind the head is a row of adapted dorsal fin rays, which serve as lures that dangle in front of its mouth, to attract prey. Sedentary by nature, the Anglerfish has two very strong ventral fins, which have been adapted into feet that can thrust the fish rapidly off the seabed in either fight, flight or when attacking prey. When lying in wait, the fish lures other prey species into its catchment area, then with a combination of thrust and opening its mouth wide, sucks seawater into the cavity along with the unfortunate prey.

Scorpionfish and Lionfish

Family
Scorpaenidae

SCORPIONFISH AND LIONFISH are rather stout fish, bottom-dwelling with no swim bladders for flotation. All are armed with venomous dorsal fins and have sharp spikes to their gill covers. They usually have warty protuberances to help provide camouflage, and are opportunistic predators. They sit still near aggregations of small fish, lying in wait for them to swim close enough to catch. The number of species now found in the Mediterranean has doubled due to Lessepsian migrants from the Red Sea.

2 STONEFISH

Synanceia verrucose

This famously venomous fish is now found from Egypt to Israel and Lebanon. Growing to around 39cm (15in), it is incredibly well camouflaged with a mottled warty appearance, lots of different colours, and often algae growing on it due to its sedentary nature. It has an upturned jaw, and spines along the back which produce a venom that is known to be fatal.

3 CARPET FLATHEAD or CROCODILEFISH

Papilloculiceps longiceps

Probably the most recognisable of the Red Sea scorpionfish that has passed through the Suez Canal, this species is now well at home in the eastern Mediterranean from Egypt up to Turkey. Found amidst stony rubble or around sponges and soft corals, it assumes the colour of its background and virtually disappears. It can grow to around 35cm (13.8 inches).

4 COMMON LIONFISH

Pterois volitans

Now also found in the Caribbean, the Common Lionfish is a voracious predator and may be found in all types of positions (including upside-down) in caves around big schools of glassy sweepers. Spreading throughout the Mediterranean, due to the widening of the Suez Canal (and having virtually no predators), it is found from Cyprus and the Greek Islands and Malta and is moving steadily westwards. It grows to around 40cm (15.75in), and the spines on the ends of its dorsal and caudal fins deliver a painfully venomous sting, as I know from personal experience!

1 SMALL ROCKFISH
Scorpaena notata

This small species only grows to 20cm (8in) and is fairly free of any facial protrusions. It is variably bright red or orange with yellowish and brown blotches and shows a black mark on the dorsal fin when it displays. It prefers low light conditions and is more often seen at night. It likes the entrances to caverns where it can be seen upside-down or high up on the cliff wall.

2 BROWN SCORPIONFISH
Scorpaena porcus

This stout species is slightly larger than the Small Rockfish at 30cm (1ft) and has a much more uniform colouring of dark brown, often speckled with cream or light brownish blotches. It has several large, backward-facing spines around the gill covers and there are two large, fleshy protuberances above the eyes. This species prefers more open water in well-illuminated areas such as seagrass meadows, where it lies in wait for its prey to swim by close to its mouth. Its spiny coverings are purely defensive – I have witnessed an Anglerfish catching one but then spitting it out.

3 GREAT ROCKFISH
Scorpaena scrofa

This is the largest of the Mediterranean scorpionfish and can grow to over 50cm (1ft 8in). Tan and light brown in colour, it has many fleshy appendices underneath the jaw and around the head, helping it to blend into the background for additional camouflage. It is found in most habitats, but prefers well-lit areas around seagrass meadows and amidst a jumble of algae-covered boulders.

4 LONG-SPINED BULLHEAD
Taurulus bubalis

This small scorpionfish grows to just 15cm (6in), but is usually much smaller than this. It has a very obvious mouth barbel or spine,

as well as a pair of significant spines behind the cheeks. It is multi-coloured and can change its coloration to suit its environment.

Squirrelfish
Family
Holocentridae

5 RED SEA STRIPED SQUIRRELFISH
Sargocentron rubrum

Covered in silvery red and white longitudinal stripes and with a spiky dorsal fin, this Red Sea immigrant is now very common in the eastern Mediterranean and is fished commercially.

Groupers
Family
Serranidae

THE GROUPER FAMILY has few representatives in the region and they have been hunted mercilessly for many years by fishermen. When you consider that some larger specimens can weigh several hundred kilos, you can understand the motivation to catch them. Thankfully, there are many established marine parks where the largest of the grouper are still abundant. They hunt by rushing at their victims with their mouths open, sucking in the prey. All fish start as females and change into males as they get older. This could upset the sex balance within populations if too many larger fish are caught.

6 BLACKTIP GROUPER
Epinephelus fasciatus

A new arrival from the Red Sea and now recorded from Israel and Lebanon, this small species grows to around 40cm (5.75 inches). Becoming more common in localised regions, it is particularly likely to be found on sand slopes near cliff edges. It has a light body, orange-brown lateral stripes, a darker area above the head and distinct black tips to its dorsal fin.

1 DUSKY GROUPER
Epinephelus marginatus

This grouper grows to an adult size of 1.5m (5ft) and is a particularly charming member of the Mediterranean fish community. Very popular with divers, it can be quite territorial, yet also tolerates its own species in large numbers whenever there is food available, such as feeding time in some marine parks. The Dusky Grouper is rich brown in colour, mottled with pale cream spots and blotches. Friendly and inquisitive, the larger males will approach divers out of curiosity. It can live for up to 50 years.

2 GOLDBLOTCH or STRIPED GROUPER
Epinephelus costae

A smaller species, typically growing to around 1m (3ft 3in), the Goldblotch Grouper has distinctive brown or golden horizontal stripes over a rather dull body colour. This fish enjoys shipwrecks and deeper water and becomes a fully mature female at 40cm (1ft 4in) long. It then changes into a male at 60cm (2ft), and at this size it is often targeted by spear fishermen.

3 CANINE or DOGTOOTH GROUPER
Epinephelus caninus

This species can grow to 1.5m (5ft), but in general only smaller ones are seen, around 50–75cm (1ft 8in–2ft 6in). Pale grey to tan in colour with a few blotched markings, it is quite timid and prefers deeper water and rocky reefs. The Canine Grouper is so named because of its strongly developed teeth, which it uses to catch and crush crustaceans. It is usually seen on its own and has a large territory, aggressively chasing others of the same species away.

4 COMBER
Serranus cabrilla

The Comber grows to around 30cm (1ft) and is pale with reddish-brown vertical bands and stripes. It has a well-defined jaw and is characterised by the saw-tooth edge of its anterior gill covers. This friendly species can be found in very shallow water, particularly around small rocky reefs at the shoreline. It enjoys *Posidonia* seagrass meadows as well as small rocky walls, where it darts for cover when it realises just how large divers are!

5 BLACKTAIL COMBER
Serranus atricauda

A smaller member of the family and an interloper from the Atlantic, the Blacktail Comber is most readily seen in the western Mediterranean. It grows to 35cm (1ft 2in) and is varied in colour, often with dark squarish patches and a broad horizontal line along its flanks. This is a territorial fish, viciously defending its home turf, and can be found in fairly shallow waters of less than 25m (80ft), around seagrass meadows and rocky reefs.

6 PAINTED COMBER
Serranus scriba

Many divers soon get to know this fish due to its very inquisitive nature and fairly gaudy coloration, which make it the most obvious of the *Serranus* species. It can grow to around 35cm (1ft 2in) and has distinctive markings around the face, with a broad dark stripe from the snout through the eye. The body has vertical, dark stripes of variable length and there is a pronounced blue spot on the belly. It occurs at depths as shallow as 1m (3ft 3in), and prefers a rocky reef filled with small holes where it hides, but also feeds on the other inhabitants of these holes. This small grouper also occurs on *Posidonia* beds, where it patrols the periphery in search of small crustaceans and fish fry.

Basslets

1 SWALLOWTAIL SEAPERCH

Anthias anthias

This is a wonderfully coloured anthias. It grows to around 10cm (4in) and is pink to red in colour with long pelvic fins, which are rounded and tinged with yellow. The tail is long and tapering and the adult male has an extended dorsal spine, which is bright yellow. It also has variable markings around the eyes and snout. All fish in the small schools are females except for the largest one, which is male. As with many other species of fish, when the male dies, the dominant female changes sex. It inhabits deeper caves and caverns and prefers low light. Large numbers can be found on deeper shipwrecks.

2 ORANGE or SCALEFIN ANTHIAS

Pseudanthias squamipinnis

Now with a strong foothold in the eastern Mediterranean, this Lessepsian migrant is the ubiquitous 'goldfish' of the entire Red Sea, with every reef and particularly every coral wall supporting great numbers of these beautiful fish. Growing to around 15cm (6in), it is the females that are golden-orange. The male is a lovely lilac to purplish pale brown, with a small red marking on the pectoral fin and a purple-red tail.

Cardinalfish

3 CARDINALFISH

Apogon imberbis

Well known to all divers who visit the Mediterranean, the Cardinalfish is very common on all reefs and is instantly recognisable. It is oval-shaped, grows to around 10cm (4in), and is a uniform reddish-orange colour with a vertical, dark line, which extends from the snout to the gill covers. The eyes have two horizontal stripes that delineate the facial stripe. Cardinalfish are always seen in large groups at the entrances to caves or large holes in the cliff face, where they shelter during daylight hours. At night, they disperse into open water to feed on rising plankton. After spawning, the male takes the egg mass of about 20,000 eggs into its mouth for protection. To provide ventilation, it chews the eggs carefully in its huge distended jaws. The male cannot take any nourishment at this time, until after the eggs hatch and the larvae are released. Although this protection is highly advantageous for the young, it is highly stressful on the parent fish. This small fish is very wide-ranging and is found from the shores of Lebanon to Portugal, the Azores and Cape Verde Islands.

4 INDIAN OCEAN TWO-SPOT CARDINALFISH

Cheilodpterus novemstriatus

A Red Sea migrant, this species has a silver-beige body with longitudinal dark stripes and an obvious dark spot at the base of the tail. First recorded in 2010, it is now firmly established in Israel, Lebanon and southern Turkey. It can grow to around 10cm (4in) and prefers a hard seabed, usually below 21m (70ft).

Jacks

JACKS ARE HIGHLY specialised, powerful fish, more at home in the open ocean than near the shore. All have very streamlined bodies with a narrow tail base and forked tail. The pectoral fins are scythe-shaped and most species resemble members of the tuna family, although are much slimmer. Most species are fish-hunters and are themselves an important source of food for human consumption.

1 BLUE RUNNER

Caranx crysos

This species of jack generally grows to around 35cm (1ft 2in) and has a distinctive, arched lateral line. Its reflective silvery body appears blue. It is a creature of the open ocean, rarely seen close to shore, but will frequent offshore reefs which have vertical walls. It hunts along the walls in small packs, rounding up smaller shoals of sardines and the like, before attacking them in an organised frenzy.

2 CREVALLE JACK

Caranx hippos

Growing to around 60cm (2ft), this fish rarely attains a larger size owing to overfishing of larger specimens. It is a highly prized species. Crevalle Jacks only form groups when they are young and rarely venture inshore. Adults are solitary and wide-ranging throughout the region.

3 MEDITERRANEAN AMBERJACK

Seriola carpenteri

This amberjack grows to around 45cm (1ft 6in) and, although very similar to the Great Amberjack, is distinguished by the much lighter yellowish markings on the head and along the flanks. The pectoral and anal fins are also tinged yellow. Little is known of its distribution. It feeds on squid and small fish, hunting in small packs.

4 GREAT AMBERJACK

Seriola dumerili

Growing to over 1.5m (5ft), the Great Amberjack has a distinctive flash of yellow or tan, which traverses the eye and spreads along the top of the body. It hunts in small packs and is in turn hunted by other jacks and even Barracuda. The hunting packs work quite close to the edge of the reef and around shipwrecks, rounding up schools of smelt and bogue before attacking with lightning speed.

5 PILOTFISH

Naucrates ductor

Growing to around 70cm (2ft 3in), this species is usually seen accompanying sharks, large jacks, tuna and turtles. This fish is very much at home accompanying larger hosts, where it cleans off their parasites and eats any left-over food scraps. It is silvery grey with dark blue-black transverse lines all along the body.

6 ATLANTIC BUMPER

Chloroscombrus chrysurus

The Atlantic Bumper is found as far west as the Gulf of Mexico and as far north as Bermuda and, as the name would suggest, has entered the Mediterranean Sea from the Atlantic and has been recorded there since 1977. Swimming in small groups, it is a pelagic species and can be found at all depths. Although they are primarily a saltwater game fish, juveniles have been found in brackish estuaries.

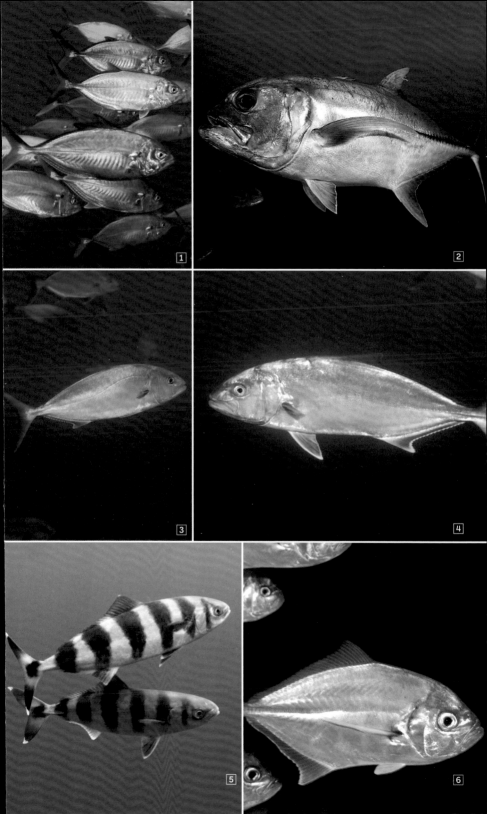

Mackerel

Family
Scombridae

1 ATLANTIC MACKEREL

Scomber scombrus

Highly prized commercially, this striking fish has a very streamlined body, compressed laterally. The first dorsal fin is sail-like and its anal fins and tail fins are deeply forked. It has large eyes and can grow as large as 60cm (1ft 10in). It hunts in small packs and is usually fished by the long-line method. It looks metallic with darker vertical and zig-zagging bands down the body.

2 INDIAN MACKEREL

Rastrelliger kanagurta

This is a relatively new migrant through the Suez Canal and is being recorded off Israel and Lebanon. It is known to form huge feeding shoals, all swimming in unison with their mouths open, scooping up plankton.

Barracuda

Family
Sphyraenidae

3 BARRACUDA

Sphyraena sphyraena

Growing over 1.60m (5ft 4in), Barracuda is regarded as one of the more successful predators in our oceans and is frequently seen in large groups near offshore reefs and rocky cliffs. It likes areas of strong moving current, particularly where diverging currents meet, carrying prey species. It becomes solitary when reaching adulthood, and it is said that these mature fish are much more dangerous. When they rest in diverging currents, Barracuda often start to circle each other, creating a great moving vortex of fish. The distribution of the species is unclear, as a smaller cousin, *Sphyraena viridensis*, is very similar to young *Sphyraena sphyraena*.

Grey Mullet

Family
Mugilidae

4 BOXLIP MULLET

Oedalechilus labeo

The Boxlip Mullet is uniform silvery grey with thin, brownish lines along the flanks. It grows to around 25cm (10in). A schooling species, it is seen in most habitats. It has large, fleshy lips with which it 'hoovers' the fine silt layers from rock surface and algae, but more commonly it is seen feeding at the surface on plankton debris and algae scum.

5 THICKLIP GREY MULLET

Chelon labrosus

With a distribution stretching from Senegal to Scandinavia, this fish is also found in all parts of the Mediterranean. The upper lip is thick and the head broad with the upper part slightly flattened. It grows to around 45cm (1ft 6in) and is uniform silvery grey with obvious longitudinal lines. It may form large groups when feeding.

Goatfish or Mullet

Family
Mullidae

6 GOLDEN GREY MULLET

Chelon auratus

The Golden Grey Mullet hunts near estuaries in small schools and is known to eat plankton, algae and invertebrates. Preferring shallower water, they can be found feeding on large plankton at the surface, as well as benthic organisms on the seabed. The juveniles move into coastal lagoons in the winter and although they are commercially fished, they are not sought after that much – rather, they are caught accidentally.

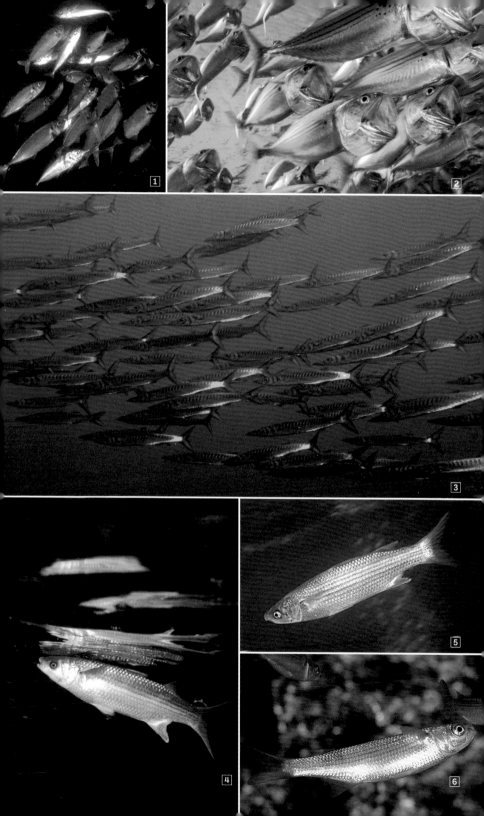

① STRIPED MULLET

Mullus surmuletus

This is the most commonly seen and easily recognised of the mullet. Sometimes referred to as Goatfish, it is variable in colour, ranging from gold or orange with red blotches, to pale cream and brown. It is also recognised by its striped dorsal fin, long, sloping head and fleshy lips. It grows to 40cm (1ft 4in). It has twin fleshy barbels, which it uses in a sweeping motion to dig in soft sand or mud in search of small worms and molluscs. Usually working in pairs or small groups, it is quite approachable and is highly prized as a food species.

② RED SEA GOATFISH

Parupeneus forsskali

This Lessepsian migrant from the Red Sea has become established in the eastern Mediterranean – for now! It is found on all dives ranging from Egypt to southern Turkey. It has an obvious near-black lateral line which is interrupted towards the tail, where a round spot can be found. It has two barbels. This goatfish is known to change colour at night from its usual creamy white to a dark red, rendering it almost invisible to predators of sleeping fish.

Gurnard

Family
Dactylopteridea

③ TUB GURNARD

Chelidonichthys lucerna

A large, robust fish with a stout, elongated head, this species is principally red or pinkish orange and grows up to 75cm (2.5ft). It has specially adapted pectoral fin rays which allow it to walk along the sea floor. This is the largest of the gurnard species. It prefers softer muddy substrates and can dive for cover should danger threaten. Juveniles and adults are found in small schools and they emit small grunting noises.

④ FLYING GURNARD

Dactylopterus volitans

The Flying Gurnard is common in some areas. Usually seen in shallow lagoons and along the reef edge, it is very distinctive when it takes off and flies over the seabed with its huge, brightly coloured 'wings'. It uses adapted ventral fin rays to 'walk' about over the seabed when searching for small molluscs to eat.

⑤ STREAKED GURNARD

Chelidonichthys lastoviza

By far the most colourful gurnard, the Streaked Gurnard is also the smallest of the species, growing to only 40cm (1ft 4in). It has a mottled coloration, tan and orange with brown blotches, and has a high triangular dorsal fin. When swimming normally, with its fin rays at its side, it resembles the Striped Mullet, but here the similarity ends. When it opens its pectoral fins, they show a brilliant blue colour.

Dragonet

Family
Callionymidae

⑥ COMMON DRAGONET

Callionymus lyra

This is a bottom-living fish. It has a roughly triangular-shaped head with a long lower jaw, and a flattened body. The male is larger, growing to 30cm (1ft), with the female just 20cm (8in). The eyes are large, slightly protruding and set on top of the head. It has two dorsal fins, with the first on the male being long and triangular and the second rayed towards the tail. During the mating season, males change colour, showing vivid blue markings around the jaw and eyes, and displaying their long colourful dorsal fin.

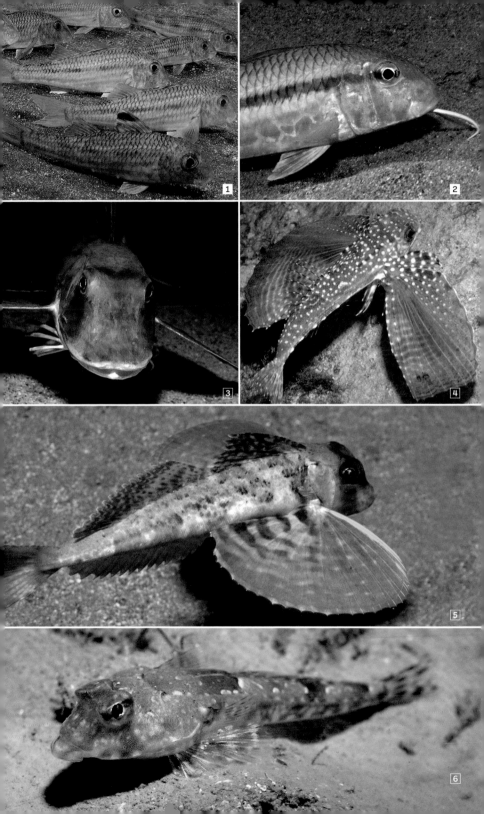

Drum

1 BROWN MEAGRE

Sciaena umbra

This fish grows to around 35cm (1ft 2in) and has a robust silvery body. The head is darker and both the pelvic and anal fins have a strong white front margin against black, making the rest of the fin look almost invisible. It usually lives in pairs. During the mating season from March to August, the male uses its swim bladder to produce a drumming or croaking noise.

Damselfish

2 CHROMIS or DAMSELFISH

Chromis chromis

This fish is frequently encountered by divers in these waters. It ranges throughout the area and prefers a rocky reef close to shore in well-lit, aerated water. It can grow to 15cm (6in) long. The adult is dull olive-green or grey with obvious scale markings. The juvenile has fluorescent blue markings on the body and prefers to hide in crevices.

3 INDO-PACIFIC SERGEANT MAJOR

Abudefduf vaigiensis

This species is becoming much more common in the eastern Mediterranean and is a migrant from the Red Sea. Often seen in large schools near the surface feeding on plankton, it can grow to around 17cm (6.7in) and has obvious dark silvery bands tinged with yellow.

Angelfish and Butterflyfish

4 YELLOW BAR ANGELFISH

Pomacanthus maculosus

Much to the delight of snorkellers and scuba divers, this brilliantly coloured angelfish is a recent immigrant from the Red Sea. Now firmly established in the eastern Mediterranean, its vivid blue overall colour is interrupted with an irregular yellow blotch, quite unmistakable from any other species in the Mediterranean.

5 ORANGEHEAD BUTTERFLYFISH

Chaetodon larvatus

Growing to around 12cm (4.7in), this new colonist is found mainly in mating pairs amongst small coral growths, where it feeds on polyps. It has an orange face, chevron stripes along a pale grey body, and black at the rear of the fins and tail. A new and interesting find in the eastern Mediterranean, it has become firmly established in the last few years.

6 RED SEA BANNERFISH

Heniochus intermedius

This is an active species, usually seen in pairs, and tends to hang out near caverns or large boulders during the day, amongst other butterflyfish and angelfish. It has a very obvious snout and raised 'eyebrows', with a black face, yellow body and dark diagonal band running from under the tail to the start of its white dorsal fin. Growing to around 20.5cm (8in), this is another new arrival through the Suez Canal, which is delighting tourists on the one hand and scaring ecologists on the other!

Picarel

1 MENDOLE

Spicara maena

Quite similar to the Picarel, this species is more rounded in shape with a more silvery body, but has the same dark blotch along its flanks. Growing slightly larger than both the Picarel and Garizzo, it can often be found in large schools over seagrass meadows.

2 PICAREL

Spicara maena

A distinctive small fish growing to 25cm (10in). It has quite a high body and is silver with a turquoise top to the flanks. It has a distinctive dark blotch midway down its lateral line. It can occur in large schools around coastal rocky habitats, but at night it rests on the seabed where it makes an easy subject for photography.

3 GARIZZO

Spicara flexuosa

Very similar to the previous two species, with the distinctive dark blotch midway along its flanks, this species is much more colourful. The belly is silvery and there are usually dark vertical bands above. Adult males exhibit brilliant blue colours on the dorsal fin and various brilliant blue lines along the flanks. The flank blotch disappears at night when at rest.

Smelt

4 SAND SMELT

Atherina hepsetus

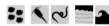

This species grows to 10cm (4in) and occurs in huge coastal schools, alongside Chromis and Bogue. A silvery colour with a bluish hue to the top of the body and a cream midrift, it has a single lateral line

down the centre of the narrow tapered body and two short dorsal fins. Fishermen catch it in large numbers, by lowering large basket nets into the water from the shore.

5 RED SEA HARDYHEAD SILVERSIDE

Antherinomorus forsskali

Growing to around 10cm (4in) and forming large schools in caverns or under large overhanging rocky ledges, this schooling fish has migrated through the Suez Canal from the Red Sea and is becoming quite common in the eastern Mediterranean. It is translucent to olive green in colour.

Bream

THE BREAM ARE PROBABLY the most distinctive of the large families of fish encountered by divers and snorkellers in the Mediterranean. There are at least 20 species to be found. The 12 most common are listed here as they are the species you may encounter around the coastline and islands. Most are deep-bodied and are strong swimmers, always active, and are covered in fairly visible scales. The colouring is virtually always metallic and the various species are recognised by the combination of blotches, stripes and bands – or the lack of them. Most are found in small shoals in mid water or near reefs, with only older specimens encountered singly.

6 BOGUE

Boops boops

This is a long, silvery fish with a metallic blue upper body and obvious lateral line. Cylindrical in shape, it can grow to 35cm (1ft 2in). It has a fine, tapering tail with yellowish, longitudinal lines on the flanks which are not always obvious. It occurs in large feeding schools on shallow reefs and near the surface, often mixed with Sand Smelt and Chromis as all three have a similar diet.

1 COMMON DENTEX
Dentex dentex

This is a large fish, growing to around 1m (3ft 3in), and has a muscular body, compressed laterally. It has a large head and steep forehead, which is slightly darker than the rest of the metallic silvery-grey body. It has obvious, silvery lateral lines to the body and a long dorsal fin. Common Dentex is usually solitary and is a voracious hunter of small fish around shallow reefs, rocky walls and wrecks.

2 BUMP HEAD BREAM
Dentex gibbosus

Although it can grow to around 1m (3ft 3in) in length, most specimens around coastal waters are less than half that size, due to overfishing. Slightly longer than Common Dentex and with a more tapered body, it is characterised by a few pale, dotted yellow to golden lines, which run horizontally along its midline. It inhabits coastal reefs and hunts near seagrass meadows, and is usually seen on its own. Unusually, the young of the species are always male. Adult females develop the hump on the front of the head.

3 ANNULAR SEA BREAM
Diplodus annularis

Growing to 24cm (10in), this silvery species is quite circular in shape with a dark band around the tail, and wide vertical, dull shading just behind the eye from the top of the head to the gill covers. This fish enjoys coastal reefs and can be found in small groups near sandy bottoms or rocky reefs. The young inhabit brackish waters. Often mistaken for White Bream.

4 ZEBRA BREAM
Diplodus cervinus

A large bream, up to 70cm (2ft), with four or five distinctive olive-green, vertical bands down the plump body. It has a high arched back and steep forehead with a narrow joint at the tail. It is usually seen on its own around rocky reefs and near seagrass meadows.

5 WHITE BREAM
Diplodus sargus sargus

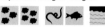

This bream is very similar to the Annular Sea Bream, with which it is often confused. It has a black spot at the body side of the tail joint and there are about seven faint, dark, vertical stripes down the body. It grows to 40cm (1ft 4in) and has a high convex forehead with a thick-lipped mouth. It occurs in large shoals and is territorial on rocky bottoms or near deep crevices at the seabed. It eats molluscs, crustaceans and sea urchins, deploying its strong jaws and heavy incisors and large crushing molars.

6 TWO-BANDED BREAM
Diplodus vulgaris

The most commonly recognised of the bream, this fish grows to 45cm (1ft 6in). It has two distinctive, vertical, black bands, one behind the eyes, the other before the tail. It has a large head and prominent eyes and the mid-section of the broad flanks has golden-yellow lines. This fish loves seagrass meadows and algae-covered rocks and always occurs in large feeding shoals. It often associates with other species of bream and is unafraid of divers.

1 SHARPNOSED SEABREAM

Diplodus puntazzo

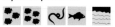

This species occurs throughout the Mediterranean, Black Sea, Atlantic coast of Spain, south along the African coast, and around the Canary Islands. As the name would suggest, it has a pointed snout, as well as pointed lips. It has a deep compressed body and usually shows pale vertical stripes, similar to the White Bream. It is a much more solitary fish, and grows to around 45cm (1ft 6in).

2 STRIPED BREAM

Lithognathus mormyrus

The Striped Bream grows to around 55cm (1ft 10in) and is widely distributed throughout all regions. Silver in colour, it has 13–15 distinct, vertical, light gold lines and an obvious, high arched lateral line. This bottom-feeder digs into soft sand looking for worms and small invertebrates and is as at home in brackish water as it is in the sea. All the fish are male at first and then change to female with maturity.

3 COW BREAM or SAUPE

Sarpa salpa

A large bream growing to 50cm (1ft 8in), it is elongate and instantly recognised by the bright golden, horizontal bands on its flanks and fleshy lips. This delightful species is always seen grazing in shoals over algae-covered reefs, close to shore. It always appears in large numbers and is constantly on the move.

4 SADDLED BREAM

Oblada melanura

This species is much longer than the others in the family, growing to 30cm (1ft). It has a large head and prominent eyes, and shows a characteristic dark blotch at the tail base, bordered by a white band on either side. It occurs in large shoals, always facing into the current, where it picks at passing plankton.

5 PANDORA

Pagellus erythrinus

Commonly found in central and western regions, the Pandora is light tan to light brown and may have some pink and bluish spots on maturity. It grows to over 35cm (1ft 2in) and is elongate with a tapering tail and rather small head. It occurs on the lower reef, beginning life as female and changing to male at about 16cm (6.5in) in size. It feeds on small crustaceans and molluscs, and has large crushing molars.

Snapper and Grunt
Family
Kyphosidae and Lethrinidae

6 BERMUDA CHUB

Kyphosus sectatrix

Much more common in the Atlantic and Caribbean, this species has gradually made its way through Bermuda and on past Gibraltar, and is now found sporadically in quite a diverse number of locations in the Mediterranean. This silvery blue, curious-natured fish grows to over 45cm (1ft 6in). It likes being close to shore and around rocky reefs, and although it is often seen free-swimming, it actually feeds on the bottom.

7 BIGEYE EMPEROR (JUVENILE)

Monotaxis grandoculis

This Red Sea immigrant is becoming more established in the eastern Mediterranean and has been recorded as far north as Antalya in Turkey. It has quite a conical head, and large eyes with a dark vertical stripe through them. In juveniles, there is a dark spot to the rear of the dorsal fin and alternating grey and white bands down the flanks. The adult is uniform silver and usually reaches around 45cm (1ft 6in) at maturity.

Triggerfish
Family
Balistoidae

1 GREY TRIGGERFISH

Balistes capriscus

The body is quite round in shape and laterally depressed, and the fins are placed far back on the body, which is crescent shaped. They lay their eggs in a sandy nest (like other triggerfish). They are commonly found around shipwrecks and under open piers, where small groups will hang out together – appearing to enjoy the comfort of shade. They can grow up to 60cm (24in), but are usually much smaller.

Rabbitfish
Family
Siganidae

2 RED SEA RABBITFISH

Siganus rivulatus

Although sometimes found in schools of 100 or more, at night this fish is always on its own and seems to pretend to be dead! During the day, it usually swims alone. It is pale green to silver and grows to around 30cm (12in).

3 STARRY RABBITFISH

Siganus stellatus

Larger than the Red Sea Rabbitfish, at up to 40cm (15.7 inches) this rabbitfish often swims with other species as a means of protection. It is frequently found in association with Coronetfish as they search the rocky reefs for food. It has a greenish-blue body spotted with yellow or brown spots.

John Dory
Family
Zeidae

4 JOHN DORY

Zeus faber

The John Dory is often referred to as the St Pierre or Peter's Fish, as the thumbprint-like marking on its side is said to have been made when Peter presented the fish to Jesus. This eyespot is known to confuse its prey. When swimming towards you, it appears an incredibly thin fish, with long, spiny dorsal fin rays. The eyes are forward-facing, providing binocular vision. This slow-swimming fish is usually easy to approach, but it is rarely seen.

Batfish
Family
Ephippidae

5 LONGFIN BATFISH

Platax tiera

This is an Indo-Pacific species, which has crossed through from the Red Sea and is happily extending its range throughout the eastern Mediterranean. Disc-shaped, it has very large dorsal and anal fins which are extremely long in juveniles. It also has a large, dark pelvic blotch just behind the pelvic fin. It is usually found in small groups, feeding on plankton in mid water, but always near reefs. It can grow as large as 60cm (24in).

Sweepers
Family
Pempheridae

6 RED SEA SWEEPER

Pempheris rhomboidea

Coming from the Red Sea, the Red Sea Sweeper has been well established in the eastern Mediterranean since 1979. Somewhat hatchet-shaped, it is usually found in shallow caves and hunts at night. It is tan-coloured and grows to around 15cm (6in).

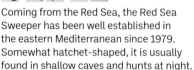

Coronetfish
Family
Fistularidae

1 CORONETFISH

Fistularia commersonii

Growing to around 1.5m (5ft), this long slender fish is usually on its own, but may accompany other fish such as jacks or rabbitfish, where it acts as a shadow, picking up food scraps on its way. It changes colour regularly, from brilliant yellow through silver to dark orange-red.

Sunfish
Family
Molidae

2 OCEAN SUNFISH

Mola mola

The Ocean Sunfish is one of the largest bony fishes in the oceans. Its body is very compressed in the vertical plane – unlike a flatfish it swims vertically rather than lying on the seabed. The dorsal and anal fins are paddle-shaped and set far back on the body, and the tail has a wavy edge. It is typically found offshore, but I have encountered it in Villefranche Bay in the south of France. Larger specimens can grow to over 3m (10ft).

Needlefish
Family
Belonidae

3 RED SEA HOUNDFISH

Tylosurus choram

This pelagic species can reach lengths of 130cm (4ft 3in) and enjoys coastal waters, particularly over the reef crest where it feeds on juvenile fish. Known to congregate in small groups in the evenings, this is a fairly new addition to the fauna of the eastern Mediterranean.

Snipefish
Family
Macrohamphodidae

4 SNIPEFISH

Macrohamphosus scolopax

This is a curiously shaped fish, unmistakable with its compressed body and elongated snout. Swimming in small groups, snout downwards, it hunts for small crustaceans, worms and fish fry around the deeper coral reefs. Generally inhabiting muddy or soft sandy areas, it is often associated with sea pens. The Snipefish has a tiny mouth at the end of a long snout, in front of large conspicuous eyes. It is usually only found in depths greater than about 25m (80ft).

Filefish and Pufferfish
Family
Monacanthidae

5 SCRAWLED FILEFISH

Aluterus scriptus

This filefish is circumtropical and is just as at home in the Caribbean. Quite shy and often seen in pairs, it is elongate and a drab grey-green in colour with very obvious scrawled blue markings all over the face and body, which become spots just before the tail. It can reach around 1m (3ft 3in) long.

6 RETICULATED FILEFISH or LEATHERJACKET

Stephanolepis diaspros

I first photographed a Reticulated Filefish in Xlendi Bay in Gozo, on a night dive in 2004 – my first sighting there of this immigrant from the Red Sea. It has a single long spine above the eyes, has a variegated coloration and can change colours to suit its environment. It can grow to around 20cm (8in), but is normally much smaller.

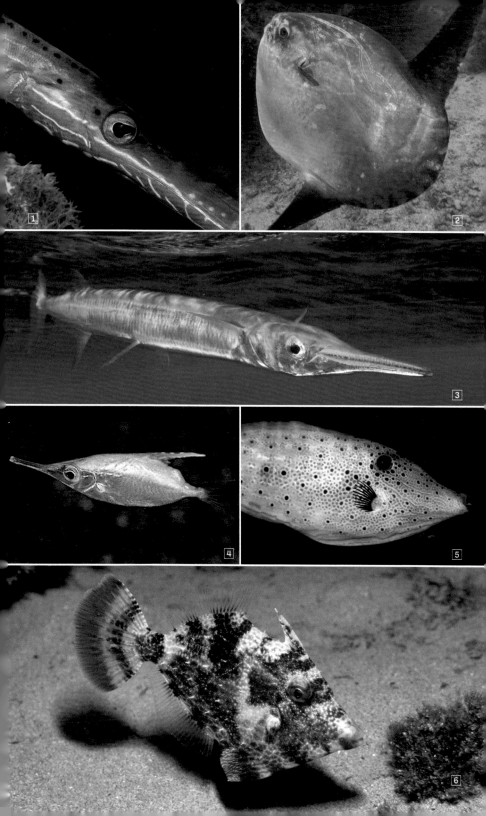

1 GUINEAN PUFFERFISH

Sphoeroides marmoratus

More at home around the Azores, Madeira and Cape Verde, this species is making its presence known in the western Mediterranean and has migrated up towards the Balearics. It is elongate with multiple pale markings, and prefers a rocky substrate.

2 YELLOW BOXFISH

Ostracion cubicus

A recent migrant through the Suez Canal, this boxfish has various colour stages throughout its life. It begins life as a tiny, yellow, square fish with black spots. Gradually it becomes more elongated, changes colour to olive-green with blueish spots, then finally becomes blue in colour with much larger patterns of darker blue spots and yellow coloration just before its wide tail. It can grow up to 45cm (17.7in).

3 WHITE-SPOTTED PUFFER

Arothron hispidus

Now recorded off Sicily, it is now apparent that the earliest Mediterranean records of this pufferfish were from before the Suez Canal was opened, indicating that it is much more widely distributed than thought earlier. Fairly rotund, its body is covered in prominent white spots with small raised spicules. It occurs in *Halophila* meadows.

Cobia
Family
Rachycentridae

4 COBIA

Rachycentron canadum

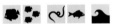

This open-water fish is found throughout the tropical Atlantic and eastern Mediterranean. It looks remarkably like the Striped Suckerfish, but has no suction disc. Its fins are quite elevated and it is rather stout in proportion.

Clingfish and Suckerfish
Family
Gobiesocidae

5 CONNEMARA CLINGFISH

Lepadogaster candolii

Found extensively from the North Sea and eastern Atlantic through most of the Mediterranean, this fish resembles a large flattened tadpole. It has distinctive light markings between the eyes, set near the front of its triangular head. It has a double ventral sucker disc with which it adheres to rocks and larger weeds.

6 CORNISH SUCKER

Lepadogaster pupurea

This fish is flattened sideways. Its head slopes gently into a long, duck-billed snout. Its dorsal and anal fins are attached to the tail and it has bluish spots or blotched markings behind its head. Variable in colour, it is found throughout the North Sea and western Atlantic. It grows to around 7.5cm (3in).

Remora
Family
Echeneidae

7 STRIPED SUCKERFISH or REMORA

Echeneis naucrates

Growing up to 80cm (31.5 inches), this is the most abundant 'sharksucker' and can be found with any and all large fish, mammals and turtles in the Mediterranean and all tropical seas.

Wrasse

WRASSE FORM A LARGE and diverse family and occur throughout the Mediterranean in large numbers. Most species are carnivorous, living on small crustaceans and molluscs, and a few have very strong jaws with which to crush sea urchins. They are found around all coastlines, quite often in very large numbers. Their common characteristics include: body length around three times their height; large, fleshy, protruding lips; prominent eyes set high on the head; one single low dorsal fin; and short, rounded pectoral fins. Juveniles of certain species may act as cleaners to other fish. Most go through various colour changes as they reach maturity and are often difficult to identify.

1 RAINBOW WRASSE

Coris julis

A distinctive, long, thin species, the Rainbow Wrasse grows to 25cm (10in) and goes through several colour changes in its life, most of which feature horizontal stripes of many different colours, from yellow to vivid orange. It may also have zigzag bands along the flanks. It has a pointed snout and tapering tail. This common species feeds low over the reef and will often follow divers, picking on any disturbed crustaceans and worms.

2 GOLDSINNY

Ctenolabrus rupestris

A small wrasse, growing to 15cm (6in), this species is light brown to golden brown with a distinctive black blotch on the upper base of the tail. It has prominent buck teeth and widely set protruding eyes. It is widely distributed amongst the algae close to the shore, where it hunts in small groups.

3 BALLAN WRASSE

Labrus bergylta

This is a large brown wrasse with mottled brownish colouring. It grows to 45cm (1ft 6in). It has a high convex head with steeply sloping brow, and large, protruding, fleshy lips. It has an intricate mesh-like pattern around the head and lower jaw. Common around coastal areas and near caverns, this fish shows no obvious differences with age, or between the sexes.

4 CUCKOO WRASSE (MALE and FEMALE)

Labrus mixtus

One of the most colourful wrasse, this species grows to 35cm (1ft 2in) and is quite common around coastal areas where it prefers a rocky substrate covered with algae and many nooks and crannies to hide in. At the male stage, it has brilliant blue markings around the head and flanks, changing to orange-red and then back to blue at the tail. The female is golden brown, with three dark markings surrounded by white at the base of the dorsal fin.

5 BROWN WRASSE

Labrus merula

This is a slender but robust fish growing to 45cm (1ft 6in). It has a moderately sized head and quite small eyes, pointed snout and thick lips. Brown to olive green in colour, its scale markings are evident, and during courtship the male will develop some blue spots and blue fin edges. This is quite a common, solitary fish found around rocky bottoms and near seagrass, where it picks around the stones looking for food scraps.

1

2

3

♂ 4

♀ 4

5

1 GREEN WRASSE

Labrus viridus

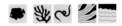

A small wrasse, olive to golden green in colour and 15cm (6in) in length. There is some controversy over this fish's classification, as it may actually be a juvenile form of another wrasse species, perhaps the Brown Wrasse. As yet, however, the link has not been found, and this could only happen in captivity over a long period. This species likes seagrass meadows and small stony areas. It is quite timid and will quickly hide under stones if threatened.

2 CORKWING WRASSE

Symphodus melops

This species has a darker blotch across the forehead and a very varied coloration, with vertical bands of golden brown and cream. It has a high dorsal fin, and shows dark bands at the tail and radiating blue stripes at the head and lower jaw. It grows to 30cm (1ft). A migrant from the Atlantic, it lives in shallow inshore waters and is relatively uncommon, but is most frequent in the western Mediterranean.

3 AXILLARY WRASSE

Symphodus mediterraneus

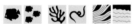

This species has a dark blotch above the lateral line at the base of the tail. It has an arched lateral line, a large head with pointed snout and prominent lips, and grows to 20cm (8in). There is a distinct yellow flash at the pectoral fins, around a dark blotch where the fin meets the body. Adult males have radiating blue lines and blue markings on the fins. It prefers the zone between the sandy area and seagrass meadows.

4 LONGSNOUT WRASSE

Symphodus rostratus

This very distinctive species grows to 8–15cm (3–6in) and has a concave forehead and a long, pointed snout. Variably speckled in colour, it also shows a white line running from its snout along the top of the back, at the base of the dorsal fins. It often swims in a head-down posture as it searches the algae for small shrimps. However, when being cleaned by other wrasse species, it hovers in a head-up position.

5 PEACOCK WRASSE

Symphodus tinca

This species has a prominent snout with white, fleshy lips and a dark brown band on the upper jaw, radiating to the eyes. The body is speckled light tan, with two or three rows of horizontal, brownish lines along the upperside. It has an obvious pinkish patch on the gill covers. It grows to 35cm (1ft 2in). This is a very common species around seagrass meadows and all algae-covered, rocky reefs, where it feeds on small crustaceans, molluscs and worms.

6 ORNATE WRASSE

Thalassoma pavo

One of the most attractively patterned wrasse, this species grows to 20cm (8in) and has various colour changes as it reaches maturity. Blue and olive green in colour, juveniles have a dark blotch midway down the back under the dorsal fin. It loves shallow, warm coastal waters where it feeds actively on small worms and crustaceans. It is always picking around the reef and, whilst it is an active predator during the day, by night it sleeps hidden from view, buried in sand.

7 PEARLY RAZORFISH

Xyrichthys novacula

This curious, slim fish with its large, rounded head and pearly sheen to its scales is found in shallow coastal lagoons and sandy inlets, away from any major water movement. Growing to 20cm (8in), it can have a light tan sheen to the scales and some radiating blue lines on the head and underside of the tail and pelvic fins on the adult males. It can dive head first into substrate to escape predators.

Parrotfish

PARROTFISH ARE DISTANT cousins of wrasse. While wrasse have fleshy lips, parrotfish have a strong, calcified beak and wide molar plate. This enables them to bite chunks of coral and grind them up to get at the soft polyps within. Important contributors to the production of sand around offshore reefs, parrotfish are voracious eaters and are always found in small groups of perhaps 10–20 females with one large 'super male'.

1 MEDITERRANEAN PARROTFISH (MALE and FEMALE)
Sparisoma cretense

The Mediterranean Parrotfish is always a delightful find. Uncommon in cooler western waters, it is more common east of the Maltese archipelago; however, it is also recorded from Madeira and the Canary Islands. The female is a brilliant mixed colour comprising a yellow ring around the eye, red snout with yellow band behind, grey head and red for the rest of the body and tail. It has another yellow patch before the tail. The males are a dull greyish blue in colour and only change sex when another predominant male has died. It grows to 50cm (1ft 8in).

2 BLUE-BARRED PARROTFISH
Scarus ghobban

Growing to around 70cm (2ft 3in), this Lessepsian migrant enjoys all types of habitat including silty lagoons and even sea grass beds and comes in varying shades of blue, aquamarine and green with pink dental plates. Now quite common in the eastern Mediterranean, it can be found readily at night when it builds itself a protective cocoon to sleep in and deter predators.

Blennies

THIS IS A LARGE family of small fish which live around coastal waters and share common characteristics. They like rocky crevices and tend to perch near the entrances of them. Most have fleshy appendages above their high set eyes and, although they resemble gobies, they have shorter blunt heads, prominent lips, and pectoral fins that function more like feet. They have single or triple dorsal fins, and lie in a flexed position. They are curious fish and can be approached quite easily, as long as you take it slowly.

3 TOMPOT BLENNY
Parablennius gattorugine

One of the largest and gaudiest blennies, this species has a thickset body that tapers to a slender tail. The forehead is high and convex and it has two very conspicuous appendages above the eyes, resembling fleshy antlers. It grows to 25cm (10in) and has seven wide, dark brown, vertical bands on the body, which varies from brown to tan and red. This wide-ranging species loves to hide in small holes, with only its head protruding.

4 PIXELATED BLENNY
Parablennius pilicornis

This species grows to 12cm (5in) and can exhibit very similar colours and markings to that of the Striped Blenny (see page 82). However, the Pixelated Blenny has two horizontal bands, with the second band running along the base of the dorsal fin. It has variable brown vertical markings and short fleshy 'horns' above the eyes. This fish enjoys algae-covered rocks and often hides in old worm holes or empty shells.

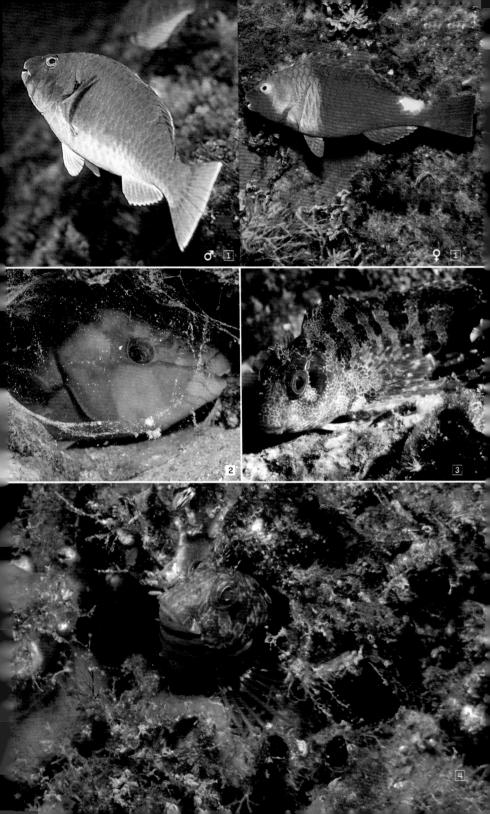

♂ 1
♀ 1
2
3
4

1 STRIPED BLENNY
Parablennius rouxi

The Striped Blenny grows to 8cm (3in) and has a single, broad, dark stripe along the body, running from the blunt forehead, across the eyes, and continuing to the tail base. It has a fairly white body with creamy spots above the longitudinal line. Its 'horns' are quite long, and three-lobed. It likes shallow, well-lit, warm waters and is quite at home in the surf zone as well as in harbour areas. It tends to hide in old worm or snail holes and is quite timid.

2 MOLLY MILLER
Scartella cristata

Lacking the obvious tentacles of some blennies, the Molly Miller has a range of short filaments on the nape. Growing over 10cm (4in) it has a mottled coloration and resembles quite a number of other blennies, but really prefers much shallower water and even rock pools.

3 STAGHORN BLENNY
Parablennius zvonimiri

When viewed from the front, this species resembles the Tompot Blenny and Striped Blenny, with its quite long, fleshy appendages above the prominent eyes. These are longer than the Striped Blenny's and not as complex as the Tompot Blenny's. It grows to 7cm (3in) and is quite common in the Adriatic and around the Greek Islands. It is only when it comes out of its hole that the brilliant white spots along the underside of the dorsal fin can be seen. This species prefers low light conditions and is more common in deeper water, as well as the entrances to caverns and under rocks.

4 TENTACLED BLENNY
Parablennius tentacularis

This blenny is very obvious because of its extremely long eye tentacles, which are rather fleshy. There are regular dark markings and bars across its flanks and back and it has a pale throat. This is a very shy species and lives in crevices.

5 DIABOLO BLENNY
Parablennius incognitus

Often confused with the Tentacled Blenny, this species has slightly branching tentacles above the eyes. The cheek has an obvious pale patch with a darker patch behind and there is an obvious notch in the dorsal fin. This is quite a small, shy blenny, growing to around 7cm (2.75in), and is mottled, with darker vertical bands.

6 CANEVA'S BLENNY
Microlipophrys canevae

This is quite a thickset fish, with stout fins that it uses to propel itself over the rocky seabed. The lower cheek is covered in small pink spots. The body has larger spots that gradually merge into longitudinal lines. A small blenny, it only grows to around 7cm (2.75in).

7 PEACOCK BLENNY
Salaria pavo

This is a curious-looking fish with a high-browed, domed head, circular blue or white markings behind the eye, and bright blue or white thin, vertical stripes and dashes along its brown and cream flanks. Larger than some other blennies, it grows to 13cm (5in) and has a slightly downturned mouth. A rarely seen species, it enjoys seagrass meadows and algae-covered rocks. It is happy in a variety of depths and is more likely to be out in the open, perched in plain sight.

1 YELLOW BLACK-FACED TRIPLEFIN

Tripterygion delasii

The appearance of the territorial, sexually mature male gives the species its name. Very obviously recognisable by its black head and vivid yellow-golden body, it has a long snout and three obvious dorsal fins. Its adapted pelvic fins are also black. It grows to 9cm (3.5in) and is widely spread throughout the region. It likes to perch at curious angles on algae-encrusted rocks and moves quickly when approached. Males are very territorial and will actively defend their territory against all other males.

2 SMALL TRIPLEFIN

Tripterygion melanurus

The Small Triplefin has three distinct dorsal fins and a black blotch surrounded by white at the tail, with a white blotch between the second and third dorsal fins. The pectoral fins are speckled white and the species grow to only 5cm (2in). This species of blenny has many different colour forms and can be quite difficult to identify. It lives amidst fuzzy, algae-covered rocks and likes very shallow, warm, well-lit water.

3 PYGMY TRIPLEFIN

Tripterygion melanurus minor

Sometimes referred to as *Tripterygion xanthosoma*, this small blenny is almost identical in nature and habit to *Tripterygion melanurus*, except for the coloration. More readily found in the Adriatic around Croatia, this subspecies appears to be spreading throughout the Mediterranean and may well be considered as a full species. It is distinguished by the lack of a distinct black spot or dash on the tail.

4 RED BLACK-FACED TRIPLEFIN

Tripterygion tripteronotum

The Red Black-faced Triplefin is common, but mainly in the northern Mediterranean from the Cote d'Azur to southern Turkey, and south around Sicily and the Maltese Islands. Very similar in shape and colours to others in the same genus, but it is a more rusty red in colour and has three distinct dorsal fins.

Stargazer
Family
Uraniscopidae

5 STARGAZER

Uranoscopus scaber

Very difficult to detect, this large, rotund fish grows to around 35cm (1ft 2in) and is often found in shallow water, where it lies buried in the sand. The top of the body is fairly flat, and has a low dorsal fin well down the back, reaching the tail, which it folds flat when hiding. It has large, upturned eyes and an even larger upturned mouth with large, overlapping teeth. It has a fleshy appendage attached to its lower lip, which it waves about to attract prey. A fairly timid fish, it will dart off in short bursts of speed if annoyed.

Weaverfish
Family
Trachinidae

6 GREATER WEAVERFISH

Trachinus draco

Similar in shape to the Stargazer, the Greater Weaverfish is longer and slimmer, but uses the same techniques for catching prey. It grows to around 40cm (1ft 4in) long and has a hinged, downturned mouth. It is tan to lightish blue on the flanks, and the back of the head is mottled. The first dorsal fin has three adapted spines, with venom glands at their bases, which it uses for defence. It usually hides in the sand with only its eyes and mouth visible.

1 STREAKED or STARRY WEAVERFISH

Trachinus radiatus

This fish's markings look almost like leopard skin along the top of its back and flanks. Its sting is painful and should be treated with potassium permanganate, and/or bathing the wound in hot water, as soon as possible. Unlike the Greater Weaverfish, the Streaked always sits high on top of the sand.

Gobies

Family Gobiidae

GOBIES HAVE TWO dorsal fins and hold themselves quite rigidly when at rest. The pelvic fins are generally modified into a sucker disc, which keeps the fish securely in place whilst balanced precariously on some underhanging rock ledge. Most gobies are bottom-dwellers and, whilst they live in similar habitats, they are a wide and diversely coloured group with different habits.

2 INCOGNITO GOBY

Gobius incognitus

This goby grows to 10cm (4in) long and is pale cream with numerous dotted patterns about the head and body. It prefers to live on the sandy seabed, always near a rocky crevice. It feeds on small crustaceans and worms and is an active daytime, as well as night-time, feeder. This is one of the few species of goby in the Mediterranean which has a relationship with an anemone, in this case the Snakelocks Anemone.

3 GIANT GOBY

Gobius cobitus

The largest goby in the eastern Atlantic and Mediterranean, the Giant Goby grows to 27cm (11in) long and is found as far east as the shores of the Black Sea. It varies from grey to cream and black, and has prominent eyes and thick, fleshy lips. It feeds on algae as well as small crustaceans, worms and

molluscs, and has powerful jaws. The spread of this fish is quite remarkable as it was originally thought to have migrated into the Mediterranean from the Atlantic. It has now extended its range further and expanded from the Mediterranean into the northern Red Sea, via the Suez Canal. It tends to lie in wait in a hole, waiting for its meal to swim by. It reaches sexual maturity at 2 to 3 years of age and lives for up to 10 years. Unlike other Mediterranean gobies, it not only eats benthic invertebrates and fish, but also algae, and it is tolerant not only of light pollution but also fresh water. This adaptability is probably the reason for its success.

4 RED-LIPPED GOBY

Gobius cruenatus

This species grows to only 18cm (7in) and is instantly recognised by its red lips. It has an elongated, circular body, a large head with black sensory spots, and an unbranched tentacle over each nostril. It has an overall blotched, dark-coloured appearance. It enjoys low light conditions and prefers to stay within reach of a rocky overhang or rock to hide under, should danger threaten. It is quite easily recognised.

5 LEOPARD-SPOTTED GOBY

Thorogobius ephippiatus

This is another very distinctive goby and cannot be confused with any other species. It grows to 13cm (5in) and is covered in dark purplish blotches on a pinkish-blue background. Widely distributed, it can be found from the Black Sea to northern Norway and the Canary Islands. Quite a sociable fish, it tends to be found in pairs or small groups, their heads usually aligned back towards a rocky recess where they quickly dart for cover if disturbed or threatened. This species prefers low light conditions and a soft substrate and, consequently, it is more commonly seen in caves and caverns. It is an active predator at night, feeding on small worms and crustaceans.

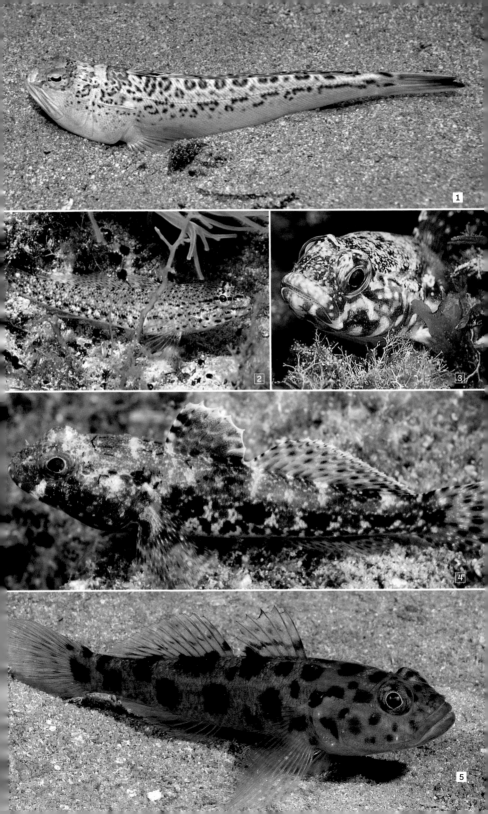

1 BLACK GOBY
Gobius niger

This is a large goby, growing to over 15cm (6in) and it quite enjoys harbours and brackish waters, occurring regularly in estuaries, coastal lagoons and sea lochs. Sexually mature at two years of age, the territorial males and juveniles are jet black (hence the name) and have a life span of five years. The Black Goby is widely distributed from Norway to the Canary Islands and Mauritania as well as through the Mediterranean into the furthest reaches of the Adriatic and the Black Sea. It is very territorial in nature and is locally common on all sandy seabeds, preferring the shelter of nearby reefs and seagrass meadows.

2 GOLDEN GOBY
Gobius auratus

This goby is quite obvious by its yellow coloration, and is found fairly exclusively in the northern Mediterranean, Adriatic and Greek Islands. Only growing to around 7cm (2.75in), it is often seen hovering in small groups above the seabed and appears to enjoy most substrates.

3 STRIPED GOBY
Gobius vitattus

Resembling the Striped Blenny in colour, it has a long dark longitudinal line that stretches from the snout through the eye to the tail. It grows to around 5cm (2in) and prefers a coralline habitat, often amongst lots of different algae, corals and sponges.

4 SAND or SLENDER GOBY
Gobius geniporus

This small species grows to 16cm (6in) and is perhaps the most common of all Mediterranean gobies. Found on the seabed on sand and around seagrass meadows, it can occur in large numbers. It has excellent camouflage, having similar markings to the speckles of the sand and gravel habitat it prefers. However, its curiosity gives it away – when you are photographing other sedentary creatures, it will come to see what is going on.

5 GROTTO GOBY
Speleogobius trigloides

This distinctively coloured small goby only grows to around 26mm (1in) and, as the name would suggest, enjoys a cave, cavern and underhanging rock habitat, well away from bright and direct sunlight. Mainly in the northern Mediterranean, it is only the red colours that have enabled me to identify this species.

6 FREI'S GOBY
Lesurigobius freisi

This particularly well-adapted goby grows to around 10cm (4in). It is only found in association with the Norway Lobster, with which it has an identical relationship to that of the partner gobies and burrowing shrimps found in more tropical waters. It darts into the orange-coloured lobster's burrow should danger threaten.

Codfish

THESE HAVE A CYLINDRICAL body and a relatively large head. Whilst many of the species hunt in large packs in open waters, the species represented here are shy creatures, preferring low light. They are soft ray-finned fish possessing fleshy barbels of some sort, have two or three dorsal fins and one or two anal fins.

① SHORE ROCKLING
Gaidropsarus vulgaris

This species is usually a uniform reddish-brown colour and has three obvious fleshy appendages, two on the snout and the other under the chin. It grows to 50cm (1ft 8in) and is a solitary fish. It prefers low light and will inhabit deeper caverns and crevices, venturing out at night to feed. Very shy, it will soon dart away should it be disturbed.

② FORKBEARD
Phycis phycis

This is a very distinctive, shy species of the cod family. It has a long barbel under the chin and an even longer one which is split into two, at the base of the underside of the jaw. It has another two protuberances at the underside of the gill flaps. It is light tan with a creamy underbelly and grows to 25cm (10in). This species is an active night hunter, feeding on small crustaceans, worms and fish fry. It is light-sensitive and always keeps well away from strong sunlight, preferring caves and deep crevices.

③ LING
Molva molva

This species resembles a very long, thin cod (also with a fleshy barbel under the chin), and grows to almost 1.8m (6ft). This is a solitary, shy fish living in caverns and crevices, but quite often in association with conger eels and squat lobsters.

Bass

④ SEA BASS
Dicentrarchus labrax

Extremely popular for the commercial market, this silvery-grey fish is common in coastal waters and grows to around 1m (3ft 3in). It is also farmed for the export market. It likes sandy, muddy or rocky coasts, where it lives in small groups. The young feed on most types of sea creatures, but adults hunt exclusively for fish. It likes offshore islands and can be observed fairly close to the surface.

Flounders

FLOUNDERS OR FLATFISH are highly specialised fish which, as they develop, become flat-bodied. They live principally on the seabed. As they grow, one eye moves around the head to join the other eye, on the side of the head which is destined to be on top. In some cases the topside pectoral fin is much longer, acting as a mating signal or sail. The body is ringed with both the dorsal and pelvic fin. All are carnivorous, feeding on molluscs, worms and crustaceans which are found in the same habitat.

⑤ WIDE-EYED FLOUNDER
Bothus podas

This flatfish grows to 45cm (1ft 6in) and is quite cylindrical in shape. Its topside skin markings are a mottled brown with circular, creamy spots and blotches. The eyes are quite wide apart and obliquely set. The right-hand side (underside or offside) is creamy white. This bottom-dwelling fish is fairly common on flat, sandy areas in lagoons and estuaries. It tends to hide under the sand with only its eyes showing, and although it is common, it is difficult to spot unless disturbed accidentally.

1 PLAICE

Pleuronectes platessa

More common in the western Mediterranean, this fish is an Atlantic migrant, with distinctive red spots on a brown upperside. It grows to 1m (3ft 3in). The eyes are raised and alongside each other, and its body is diamond-shaped. This is a commercially harvested species and much sought after in the marketplace. It frequently enters brackish water and is quite approachable.

2 TURBOT

Psetta maxima

This is one of the largest flatfish to be found in the region and ranges from the Black Sea to the Atlantic. A creature of deeper water, it may be encountered around deeper wrecks. It can grow up to 1m (3ft 3in) and has wide-set, protruding eyes. The upperside skin has no scales and is covered by rough bony tubercles. This is a bottom-dwelling species and likes company, quite often being found in small groups. It can live for more than 20 years. It spawns in July and August. The four egg sacs are always carried in the left-hand side of the body, with a large cavity on both top and bottom.

3 ECKSTRÖM'S TOPKNOT

Zeugopterus regius

Unlike many other flounders, this species prefers a rocky substrate where it feeds on small crustaceans and fish. It only grows to 20cm (8in) and has an elongated shape. It has very vivid pink and purple blotches around the head and upper body.

4 EUROPEAN FLOUNDER

Platichthys flesus

This flounder lives on fine, sandy or muddy bottoms and will frequently enter brackish water in the search for mussels, small crabs and fishes. Its growth is very rapid, depending on the abundance of food, and adults can reach over 50cm (1ft 6in). Triangle-shaped, its dorsal and anal fins spread out into a fan shape towards the tail. It is multicoloured and usually flecked with light spots over a drab background.

5 MEGRIM

Lepidorrhombus whiffiagonis

The Megrim is a large left-eyed flounder and the lower jaw is quite prominent. More round in aspect, it is rarely found inshore, but is more often seen at night when it is an active feeder.

6 BRILL

Scopthalmus rhombus

Typically flounder-shaped and resembling a turbot, it has smooth skin, lacking bony tubercles. Mottled brown above and a creamy underside, it is left-eyed and as you look at the head from above, the margin of the fin, on the right side under the eye, is segmented and very obvious, which helps identification. It can grow to more than 90cm (3ft).

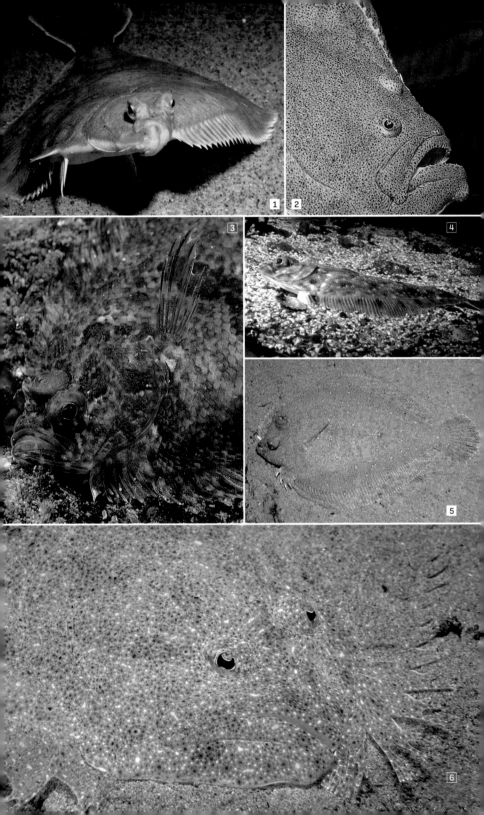

OTHER VERTEBRATES

Seals

1 MEDITERRANEAN MONK SEAL
Monachus monachus

Classed as Critically Endangered, this seal's total population is now only around 700. Resembling the monk seal species found in Hawaii, the Mediterranean Monk Seal is found in Madeira and a couple of isolated pockets off the coast of Morocco, but principally around the Greek Islands of the Adriatic. Very shy in nature, it is (quite rightly) very wary of human interference. It is quite similar in shape and size to the Common or Harbour Seal, and there may be some misidentification between the two, resulting in inaccurate observations on distribution.

2 GREY SEAL
Halichoerus grypus

Whilst this is a species more at home in the North Sea and west coasts of France, Spain and Portugal, it is a wanderer and several have been found at the western Mediterranean coast of Spain. It is quite large at over 3m (10ft) long and has a distinctive horse-like head.

Cetaceans

3 BOTTLENOSE DOLPHIN
Tursiops truncatus

This is the most commonly encountered dolphin in the region and pods can be found in many areas, such as at Gibraltar, Sicily and around the Greek Islands. It has a friendly face with a long beak and high forehead, and grows up to 4m (13ft) long. This species enjoys riding the bow wave of fast boats, and this is the way that most tourists encounter it in the Mediterranean.

4 FALSE KILLER WHALE
Pseudorca crassidens

Growing to around 6m (20ft), this cetacean is a rare but fairly frequently seen species of large toothed dolphin. It is generally black with a grey throat and neck, and is known to display similar characteristics to the true Orca, attacking and feeding on other cetaceans. It is also kept in captivity, and some of the world's wild populations (particularly around Hawaii) are studied extensively. It is more commonly pantropical, but strandings have occurred in the North Sea and it is regularly seen off Gibraltar.

1 MINKE WHALE
Balaenoptera acutorostrata

The Minke Whale is surprisingly common in the Mediterranean and tends to favour coastal locations, whether near the mainland or offshore islands, where upwellings of nutrients attract larger schools of fish.

2 SPERM WHALE
Physeter microcephalus

The Mediterranean population of this whale appears to be completely separate from Atlantic counterparts, and inhabits the deeper basins of the Mediterranean where it has a similar lifestyle and diet to its cousins, feeding mainly on squid. Often seen passing by the deep trenches close to shore, it is regularly observed in small groups on the surface. It can stay underwater for more than 20 minutes whilst searching for food, and its massive 4m (13ft) flukes can propel it into the depths quickly and efficiently.

Turtles

3 HAWKSBILL TURTLE
Eretmochelys imbricata

This is certainly an endangered species in the Mediterranean, with its nesting beaches under threat from tourist development. Often caught in trawl nets, this turtle feeds on jellyfish, sponges and some algae. It has a hawk-like beak and overlapping bony plates along its back, unlike the Green Turtle. Growing to 1.20m (4ft), the Hawksbill Turtle is more commonly seen in the western Mediterranean, where it migrates from the Atlantic. The marine museum Aula del Mar, in Malaga, cares for wounded individuals, and these can be observed by the public whilst the turtles are in care.

4 GREEN TURTLE
Chelonia mydas

Once very common in central and eastern regions of the Mediterranean, but now mainly confined to the Greek Islands and Turkish coasts, where there are still protected breeding beaches. However, many turtles are caught each year as bycatch from drift netting, and it is now considered endangered. This species has non-overlapping body plates, unlike the Hawksbill, and is smaller at up to 1m (3ft 3in) long. The other obvious difference is that there are only two plate sections on the forehead of the Green Turtle, but four distinct plate sections on the head of the Hawksbill Turtle.

5 LOGGERHEAD TURTLE
Caretta caretta

Rarely seen, this species is known to nest on Gozo in the Maltese Islands. It feeds on jellyfish as well as bottom-dwelling invertebrates, leathery corals and sponges. It has very strong, powerful jaws and a smooth shell with five raised scutes or scales down the centre-line of its back. This species is also pantropical.

INVERTEBRATES

RHIZARIA

① RED FORAMINIFERAN
Miniacina miniacea
This is an ancient yet tiny foramaniferan which only grows to about 10mm (0.3in) in height. It is multi-stemmed and each stem is tubular. It closely resembles both a tiny sponge or coral, yet is in a completely different family group. Usually missed as they are so small they are a brilliant reddish purple in colour and will form large colonies of individual structures on rock.

SPONGES

Phylum Porifera

SPONGES ARE SESSILE ANIMALS, simply constructed with a single-body cavity that bears large exhalant pores or oscula, and smaller inhalant openings lined with special cells called choanocytes, through which nutrient-rich water is passed. The exhalant tube is usually located at the highest point of the animal, to allow for waste water to be carried away more efficiently. Sponges have no internal organs. Yet despite their simplicity they are very successful organisms and are found worldwide in many different underwater habitats. The hard structure of the sponge contains small calcareous or siliceous spikes or spicules, which support the sponge. Sponges are often difficult to identify, as they change shape to suit their environment. In very exposed locations, they will be flattened, often covering large areas of rocky substrate. In calm sheltered conditions, many grow quite spectacularly with large branching arms and delicate formations.

② SPIKY SPONGE
Acanthella acuta
Relatively common, but only growing in small clumps in fairly shallow water, this sponge is often found on wrecks where it enjoys a firm holdfast. Obvious by its spiky shape, growing out over 25cm (10in), this orange sponge is variable in shape with obvious large oscula set back into the body and often has a rather fuzzy appearance.

③ YELLOW TUBE SPONGE
Verongia aerophoba
Enjoying a rocky substrate in fairly shallow water from 5–25m (17–83ft), this bright yellow sponge is tinged with symbiotic blue algae which quite often makes it look greenish or brown in colour. It forms small clumps with large, finger-like tubes approximately 3cm (1.25in) in diameter and up to 12cm (4.75in), long with an obvious exhalant opening at the end of each tube.

1 YELLOW BRANCHING SPONGE

Axinella damicornis
Often mistaken for Spiky Sponge, the Yellow Branching Sponge tends to be low-forming with irregular outcrops of exhaling tubes. Almost lemon-yellow in colour, with a slightly 'fuzzy' appearance, this small sponge is rough in texture, and it prefers shaded areas, rarely growing below 15m (50ft), and reaches only 10cm (4in) in height.

2 BRANCHING TUBE SPONGE

Axinella polypoides
Quite often found in muddy pockets on the lower rock substrate inside caverns, this columnar sponge, yellow or orange in colour, forms cylindrical tree-like branches. Depending on the surge quality of the water, it may develop many small branches, or only one or two very long branches, growing up to 1m (3ft 3in) in length. Quite fragile, it can be easily damaged. Of the three large and similar species of *Axinella*, this species is recognised by its lack of tuberculous growths and it tends to have longer, more cylindrical branches. *Axinella canabina* has many small, more-jagged, irregular branches rising vertically from a single stalk, and *A. verrucosa* is fairly smooth and cylindrical, but also is comprised of many small branches and projections. The branches of *A. polypoides* are often so long that they get too heavy to be self-supporting and drape themselves over the rocky substrate.

3 POTATO SPONGE

Chondrilla nucula
Particularly widespread in most areas of the central Mediterranean, this species is especially resistant to surge and rough seas, as well as being tolerant of mild pollution from sewage. Resembling small, brown, new potatoes, this sponge grows to a maximum size of just 2cm (0.75in) and forms small patches on hard rock. Enjoying full sunlight, it is located on the tops of rocks and in shallow water to a depth of around 7m (23ft).

4 CLATHRINA

Clathrina coriacea
Rarely found in depths of less than 10m (33ft), this golden-yellow honeycomb sponge is only 3cm (1.25in) in height, but grows extensively over large areas. It prefers shaded areas and is common in caverns and under overhangs. Constructed of many small tubules, which link and weave together, it forms a raised pad, which quite often hangs down from cave walls. It looks somewhat like a lump of small netting, with the exhalant valves or oscula at the ends of the interlocking tubes. A similar species, *Clathrina clathrus*, is white and is also found in both the Mediterranean and Atlantic. Clathrina is commonly associated with large colonies of the tunicate Dendrodoa, with which it interleaves, creating a marked colour contrast between the yellow of the sponge and the red of the tunicate.

5 BORING SPONGE

Cliona viridis
In varying shades of yellow and green, this species is probably much more common than originally thought, as its small size is so easily overlooked by marine life observers. Looking like a series of excavated holes in the top of limestone rocks, the Boring Sponge has many inhalant and exhalant holes lined with calcareous spicules to trap organic matter. This sponge bores deeply into soft rock and forms large colonies, the large holes of which are quite often used as refuge by small crustaceans or gobies.

6 OYSTER SPONGE

Crambe crambe
The Oyster Sponge is quite common and found in fairly well-lit waters from 5–30m (17–100ft). Commonly associated with Thorny Oysters (see page 132) and other sedentary bivalve molluscs, this brilliant red encrusting sponge covers the outer shell of the bivalve. Ridged by raised oscula found along the exhaling channels, the colonies grow from 10–20cm (4–8in).

1 BREADCRUMB SPONGE
Dyscidea fragilis
Forming large mats of more than 1.5m (5ft) across in low light conditions, principally under overhangs or in caverns, the Breadcrumb Sponge is uniform grey tinged with light purple. This sponge forms many raised oscula and is quite distinct from any other species.

2 BLACK SPONGE
Ircinia spinosa
Particularly common in central and eastern Mediterranean regions, the Black Sponge grows to approximately 20cm (8in) in diameter and is a low, encrusting species found on the tops of well-illuminated rocky surfaces in all depths.

3 LOBED SPONGE
Oscarella lobularis
Commonly found in caves and caverns, this fleshy demosponge has neither spicules nor flexible collagen (spongin). Forming large sheets over 10cm (4in) thick, it shies away from natural sunlight. Its large interconnected tubes are pale cream in colour with brownish tips. It can cover large areas of cliff wall or overhang and will colonise rocks and algae.

4 PINK CAVE SPONGE
Petrosia ficiformis
This common sponge is found in most cave and cavern situations and can cover large expanses of cavern wall in long, sometimes connecting, lobes. With highly visible oscula, the colours are variable, combining peach, grey-white and sometimes purple. This is a favourite food species of the nudibranch Spotted Doris (*Discodoris atromaculata*), seen opposite and on page 138.

5 ORANGE SPONGE
Spirastrella cunctatrix
One of the more common sponges found in the Mediterranean, the Orange Sponge grows over rocky walls and covers most other organisms. Bright orange in colour, with quite obvious channels that run over the body leading to the oscula, it is often mistaken for *Crambe crambe* (page 100), but grows over a much wider area, often taking over a rocky wall.

6 GREEK BATH SPONGE
Spongia officinalis
Lacking calcareous spicules, the Greek Bath Sponge has a complex structured skeleton of spongin fibres, making it extremely flexible. A light pink or uniform grey in colour, this sponge has long been collected commercially for sale as a bath sponge. The oscula are sparse but conspicuous, being slightly raised from the textured body. It is common in caves and areas of low light from shallow water to below 40m (130ft).

7 LARGE ENCRUSTING SPONGE
Suberites domuncula
This is a large sponge, developing irregular mounds and spheres with obvious oscula. It can grow to more than 1m (3ft 3in) wide and 7cm (2.75in) thick. Quite hard and leathery in texture, it has three different types of spicules in its formation. It is quite often found growing on a hermit crab's shell, where it eventually dissolves the shell, creating a new secondary home for the crab. This allows the crab to grow larger without the need to change shells.

8 PURSE SPONGE
Sycon ciliatum
A quite distinct small sponge with a single ovoid or spherical tube around 7.5cm (3in) long. It has a single large oscula at the top, surrounded by a stiff, spiky collar. Preferring shallow water and found in beds of mixed algae, this is an annual species, releasing its larvae in the spring.

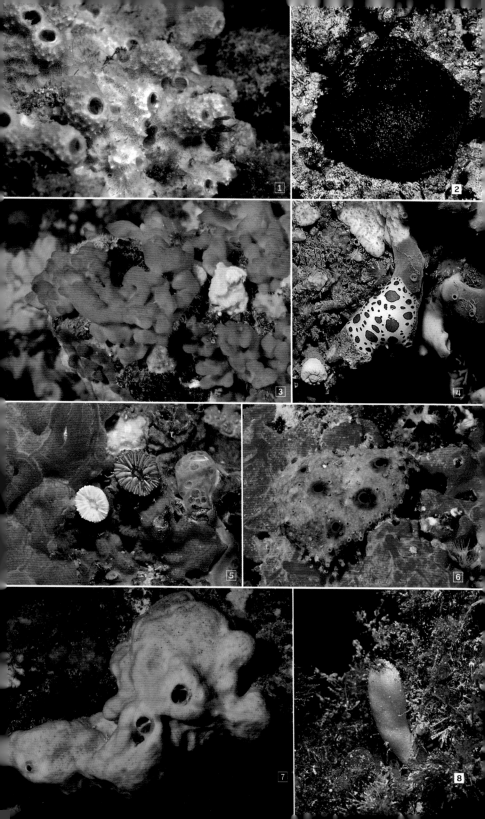

CNIDARIANS and CTENOPHORES

THIS IS A LARGE GROUP of fundamentally simple animals, widely distributed and seemingly highly diverse, with a wide array of forms that are not obviously related to one another. The phylum Cnidaria includes true hard corals, soft corals, sea pens, sea fans, anemones, sea firs (hydroids), zoanthids and jellyfish. They come in two different forms – an attached (sessile) polyp or a free-floating medusa. Ctenophores (phylum Ctenophora) are a group of unrelated but jellyfish-like animals.

All cnidarians are radially symmetrical and have a single body cavity with a single terminal opening, usually surrounded by one or numerous rings of tentacles. It is these tentacles in various forms and adaptation which unite the group. Each tentacle is armed with stinging cells used for defence or aggression for catching prey. These stinging cells may also be found on other parts of the body. (See page 22, Sea Stingers)

A typical cnidarian has a single mouth, ringed with tentacles with which the animal stings, stuns or captures its prey, to be drawn into the mouth and thence to the single, sac-like body cavity. Waste matter also exits through the mouth. Most cnidarians reproduce sexually, releasing floating larvae. Some sessile species will split or bud to produce new polyps that remain attached, thereby increasing the size of the colony, whilst others release free-floating eggs and sperm which are fertilised in open water, producing medusae which eventually settle onto the seabed.

Hydroids

COMMONLY REFERRED TO as 'sea firs', hydroids usually live attached to rocky substrates and have a rather complicated life history. They are often mistaken for plants or algae. They form delicate fern-like or feather-like groups and can be quite profuse in certain areas. During sexual reproduction, they produce a free-swimming, tiny jellyfish called a hydromedusa. Although most hydroids have tiny polyps, they pack powerful stings and contact with them should be avoided.

1 SEA FERN
Aglophenia spp.
Difficult to identify specifically, *Aglophenia* species are typical of the feather type of formation, with each 'fern leaf' reaching around 20cm (8in) in length. Unbranched, the side 'shoots' are opposite each other and form two vertical rows. The male gonophores tend to congregate along the stem, at no set intervals. Very common on the tops of rocks, they are often overlooked and can sting the softer parts of your skin.

2 SEA FIR
Eudendrium rameum
This is a short, stubby hydroid with a small, quite thick basal stem, quickly branching out. The polyp heads are arranged in alternate sequence, crowned with 24 tentacles. The male gonophores will also be clustered around the polyps. Found in fairly shallow water of less than 15m (50ft), the hydroid thicket will rarely exceed 20cm (8in).

3 SINGLE HYDROID
Corymorpha officinalis
Living under soft sand or mud, this single-stalked hydroid prefers low light, deep water and usually only appears at night. Virtually translucent, it may have faint stripes on the extended stalk and resembles a very long, thin anemone.

4 SEA FEATHER
Gymnangium montagui
These rather striking fronds of *Gymnangium* stand over 10cm (4in) tall and occur in depths of over 20m (66ft), typically on exposed cliff walls where there is plenty of aeration and passing planktonic particles. Golden-brown in colour, its colonies may contain several hundred fronds.

Jellyfish, Siphonophores and Ctenophores

JELLYFISH ARE FREE-FLOATING medusae in which the polyp stage of the life cycle has either been totally suppressed or extremely reduced. The upper surface of the jellyfish is generally smooth to the touch and is known as the aboral or exumbrellar portion. The subumbrellar portion is underneath and contains various combinations of tentacles and mouthparts, armed with a variety of stinging cells. Jellyfish are capable of directional movement by pulsing the outer bell, creating a staccato propulsion. More often than not, though, they are at the mercy of tidal movements and bad weather and can be washed up on tourist beaches in large numbers. Siphonophores or hydrozoa are jellyfish-like but form a separate group within Cnidaria, while ctenophores belong to their own distinct phylum.

1 LUMINESCENT JELLYFISH
Pelagia noctiluca
This oceanic jellyfish has a large, spotted bell of around 10cm (4in). The spots are actually warts, armed with stinging cells. It has 16 lobes, eight sense organs and eight marginal tentacles with four shorter frilled mouth tentacles under the bell. The marginal tentacles can be over 1m (3ft 3in) long, making them rather hazardous to swimmers during the summer months when this jellyfish can be quite prolific in inshore waters. There is no sessile stage in its reproduction, with the adult releasing juvenile medusae in the autumn.

2 BELL JELLYFISH
Rhizostoma pulmo
The largest of the jellyfish encountered in the Mediterranean, this species' outer bell can be over 1m (3ft 3in) in diameter. Very solid in construction, it overwinters in deeper water and migrates from the Atlantic. The eight short clubbed tentacles are subdivided into numerous frilled mouths and employ a mucus coating to trap food particles.

3 MOON JELLYFISH
Aurelia aurita
The Moon Jellyfish is one of the most common of all the jellyfish and occurs in all of our oceans and seas. Recognised by the four purplish or blue gonad rings found on the top of the bell, it can grow up to 40cm (1ft 4in) in diameter. The bell is surrounded by a ring of small tentacles. It is not harmful to humans but preys on small fish and other planktonic larvae. This jellyfish uses the sun as a compass and forms breeding aggregations in late summer, followed by extensive mutual migration into shallow coastal waters. The sedentary stage is found under rocky overhangs well aerated by tidal movement, releasing young medusae into open waters in the spring.

4 CASSIOPEIA
Cothyloriza tubercolata
Occasionally called the 'Fried Egg Jellyfish', the Cassiopeia comes into coastal waters in late summer, driven by the need to spawn in shallow, well-lit waters. With a diameter of around 40cm (1ft 4in), this large, distinctive medusa has a creamy yellow bell top which is smaller than its tentacle ring. The tentacles are short and terminate in purple spots containing the stinging cells. Juvenile Sand Smelt and Bogue are often found swimming within the 'protective' embrace of the tentacles.

5 SEA GOOSEBERRY
Ctenophore spp.
These ctenophores (a phylum of jellyfish-like animals) have rounded, cylindrical bodies, growing up to 3cm (1.2in) in diameter. They have eight distinct swimming combs, made of hair-like cilia, which extend most of the way around the body. The mouth is located at the bottom, where two feathery tentacles grow. These can reach over 14cm (5.5in) long, and are used to catch prey. They are pelagic species, often being blown inshore in the autumn months where they collect in rock pools. When seen at night, the swimming cilia are iridescent and appear to emit light as the rainbow-coloured cilia beat to aid its propulsion through the water.

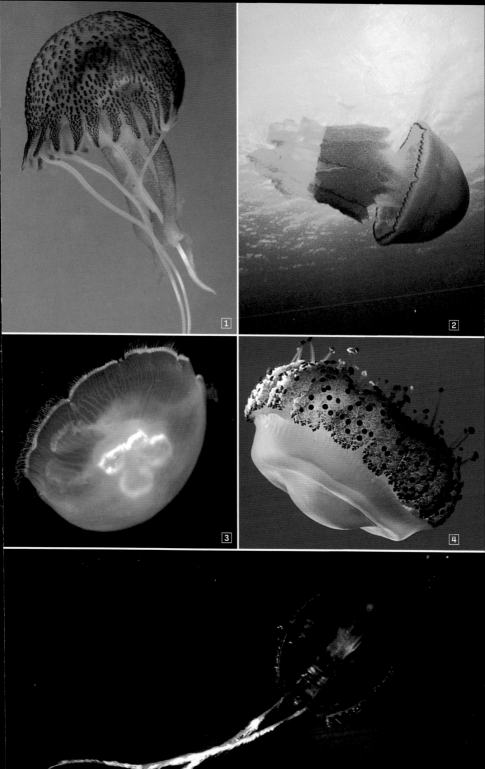

[1] COMB JELLY

Beroe cucumis
The Comb Jelly is a very simple and translucent ctenophore, with no tentacles. It mainly moves with the sea's currents but does have limited locomotion from its cilia, which beat along its slightly ribbed flanks. Flashes of light can be seen when the cilia move in unison.

[2] PORTUGUESE MAN O' WAR

Physalia physalis
This is a very distinctive siphonophore and takes its name from the obvious sail-like balloon which sits above the surface of the water. It moves by wind power and may have very long, highly toxic tentacles. Following sea storms, literally hundreds can be washed up on the shore and they must be avoided at all costs. The blue or mauve bladder or pneumatophore can grow as large as 9–30cm (3.5–11.8in) long and stand as high as 15cm (6in) above the water. The colony of zooids which it supports may have tentacles between 10–30m (33–100ft) long.

[3] HYDROID MEDUSA

Neoturris pileata
Growing to around 15mm (0.5in) long, this Hydroid Medusa is widely distributed throughout the Atlantic, North Sea and Mediterranean, even reaching as far as the Black Sea region. Large aggregations can often be found locally (likely the result of a mass spawning).

Octocorallia

THIS FAMILY INCLUDES the soft corals, sea fans and sea pens. The medusoid stage in octocorallians has been suppressed and all of the animals represented in the Mediterranean are in fact colonies consisting of loosely connected polyps. The sea fans are more complex and create a hard skeleton underlying the polyps. Whilst this makes a perfect shape for netting passing food particles, sea fans are also prone to colonisation by other species and can become completely overgrown. Sea pens tend to hide under soft sediments during the day and extend their soft bodies by infusing water and stretching upwards to feed at night.

[4] FALSE RED CORAL

Parerythropodium coralloides
This species tends to form thin encrusting sheets which colonise gorgonian sea fans. Spreading up from the base, these colonial octocorallians have a ruby-red body and lighter, widely spaced, raised polyps. They are known to form small clumps on rocks and may even overgrow shells, but they are more commonly seen on *Eunicella* sea fans where the colour change is more striking.

[5] SEA FINGERS

Alcyonium acaule
These hand-shaped colonies with stubby fingerlike projections up to 12cm (5in) long are always rosy red, with polyps which are tinged the same colour or even light yellow. This species can have quite a wide-spreading base attached to hard substrate. Closely related to Dead Men's Fingers, its compacted colonies can be found in depths of below 20m (66ft), quite often on shipwrecks.

[6] DEAD MEN'S FINGERS

Alcyonium palmatum
Compared with Sea Fingers, this species forms more slender, upright branching colonies up to 12cm (5in) tall, connected to a hard substrate by a more slender columnar base. The body colour can vary from pale yellow through orange to red, but the polyps are always white and more spread out than on other related species. All of the *Alcyonium* species are characterised by their stubby, digiform-like ramifications, always cylindrical in shape and stretching from a common base, which has little or no polyps. They are heavily preyed upon by a number of different nudibranch species. The polyps tend only to extend for feeding during a strong surge, current or at night; at other times, the 'fingers' resemble the form of a jointed red sponge.

1 ROUND SEA PEN
Veretillum cynomorium

Particularly common in the western Mediterranean around Morocco, Gibraltar and the Costa del Sol, this species prefers soft, fine sand or a muddy substrate, where its long foot can stay buried. If the habitat is threatened, perhaps in a violent storm or through pollution, it has the ability to release the water from its column, pull free of the seabed and gently roll off in the current until it finds another suitable habitat. Growing to over 20cm (8in) in length, the Round Sea Pen expands and retracts its cylindrical body by pumping water through its cavity. Usually only seen at night, this is one of the larger sea pens to be found in the Mediterranean. It is pale orange or yellow, with polyps that vary in colour from white to brown and can extend to over 2cm (0.75in).

2 COMMON SEA PEN
Virgularia mirabilis

Little seen, this small species (up to 17cm (7in) in height) prefers deeper waters with some current to spread its distinctly grouped polyps. Found in clusters of three to eight and located in two rows, the polyp and column are able to turn towards the prevailing current and create a net with which to capture food particles. This species prefers a clean, fine sandy seabed and is generally associated, although not exclusively, with sea lochs and fjords.

3 TALL SEA PEN
Funiculina quadrangularis

This is the tallest of the sea pens found in the Mediterranean and is recorded from the north-western British Isles, numerous fjords in Norwegian waters and all the way through the deeper waters of the Mediterranean. It can grow as large as 1.7m (5.5ft), and so is fairly easy to identify! Unlike the Common Sea Pen, the stem is unbranched and the polyps sit directly on the stem. Its colour is usually reddish yellow or cream. It occurs on soft bottoms below 20m (66ft).

4 PHOSPHORESCENT SEA PEN
Penneatula phosphorea

Usually pinkish in colour, this species forms a fleshy central column, with various side branches arranged on a single plane. The polyps are white and the arms only extend to about half of the length of the animal, the rest of which is hidden below the sand. When moved suddenly, the animal may flash phosphorescently. This species can grow to 30cm (12in).

Sea Fans or Gorgonians

SEA FANS ARE reasonably common throughout the Mediterranean and although there are few varieties, they are on the whole quite colourful, with the predominant species being *Paramuricea clavata* which can cover huge sections of cliff walls and wrecks. Owing to their sedentary nature, they prefer areas of moving water. Sea fans are characterised by an erect colony of individuals which form an intricate web of dividing branches with extended polyps. They are aligned along a single plane across the prevailing current, to snare prey as it passes in the tidal stream.

5 PRECIOUS or RED CORAL
Corallium rubrum

Extensively collected and highly valued since classical times, this species is directly related to other gorgonians (although it does not grow in the typical fan shape). Larger specimens have been intensively harvested over the years and are now only found in very deep water or in undiscovered caves. Each year, commercial coral divers are still killed or paralysed by the crippling 'bends' in their illegal search for greater and better-formed branches. Recognised as a protected species, the small, knobbly, three-dimensional branching colonies prefer deep caverns and other areas of very low light, including deep water. The classic red colour is instantly recognisable, with white, widely spaced polyps. This fragile coral is variable in size depending on its location. In general diving areas in caves and caverns, it only grows to 3cm (1.25in); larger fans may be found in deeper waters.

1 YELLOW SEA FAN
Eunicella cavolinii
Very similar to *Eunicella verrucosa*, except that the polyps are arranged in whorls at the tips of each branch. It enjoys vertical cliff walls where the fans, growing to around 40cm (1ft 4in) lie across the current to collect food particles. Found in shallow depths from 10–30m (33–100ft).

2 WARTY SEA FAN
Eunicella verrucosa
This species is closely related to *E. singularis*, but is more richly branched, forming a wider fan shape, more like a true gorgonian, up to 30cm (12in). Protruding out from the branches, the polyps are white and have no pattern. They are arranged in double rows towards the tips. This distinctive species is found as far north as the British Isles and is always a delight for divers to find there, as the thought of exotic corals in British waters is something of a novelty! Preferring more sheltered locations and deeper water than *E. singularis*, this species is eaten by various nudibranchs.

3 WHITE SEA FAN
Eunicella singularis
The most common of the sea fans and ranging throughout the Mediterranean, the White Sea Fan lives on rocks or wrecks alike in depths of below 12m (40ft) and can cover large areas. From a common stem, the branches are elongated and generally stretch upwards to a height of 30cm (12in). It hosts a microscopic, symbiotic alga, which may give it a slightly green coloration. The small polyps are darker than the body.

4 RED SEA FAN
Paramuricea clavata
The most common and colourful of all the Mediterranean sea fans, the Red Sea Fan adorns most vertical cliff faces in the central and western areas, growing up to 60cm (2ft). Depending on certain conditions, the fans will vary in colour from white through yellow to ruby red and purple, quite often on the same animal. Several areas are particularly well known for their extensive colonies, such as the Medas Islands, the Straits of Bonifacio between Corsica and Sardinia, the islands of Elba and Ustica, and the wrecks of Gibraltar. The Red Sea Fan particularly likes clear, clean water on offshore rocky seamounts where it gets the most benefit from coastal currents. The fans are also an important habitat for several species of nudibranch and mollusc, and even catsharks lay their egg cases on them where the aerated water is most beneficial to the developing embryo.

5 GORGONIAN
Lophogorgia sarmentosa
Rosy red in colour, this sea fan has widely spaced branches which spread out from a common root stalk. The polyps are slightly paler than the body and create an effective net to capture food particles. Generally found in depths greater than 20m (66ft), this sea fan is quite rare in most areas.

Hexacorallia (Anemones)

THIS SUBDIVISION OF THE cnidarians is the largest group and includes the anemones, zoanthids and true corals. The name indicates that the polyp tentacles are arranged in groups of six rather than eight (octocorallians such as Dead Men's Fingers or sea fans). Most anemones are solitary and attached to a hard substrate by means of a basal sucker. Others burrow into soft sand or mud, creating a protective tube into which they can withdraw. Some species enjoy living in close proximity due to the very nature of their reproduction, such as the Snakelocks Anemone which multiplies by fission, creating two complete adults. Others produce live young. Some use pedal laceration, where part of the column or base is detached and grows into a new polyp. All anemones are carnivorous, trapping unwary small fish or invertebrates within the sticky grasp of their tentacles. Many creatures are also seemingly immune from the stinging cells and live within the range of the tentacles for protection from other predators. Two species of hermit crab are even more closely associated with anemones, playing direct host and gaining additional protection for themselves.

1 BEADLET ANEMONE
Actinia equina
This species is one of the more common of the shallow water anemones found in the western Mediterranean region and may be left exposed at low tide or found in shallow rock pools. As children we called it 'bloodsucker' as its sticky tentacles could grasp a finger, as well as, of course, for its vivid red coloration. It can grow to around 6cm (2.3in) tall and the same in diameter. The column is smooth and the oral disc has 192 tentacles in five or six rows, which can be quickly retracted. Below the disc and sometimes formed as part of the collar there are 24 hollow warts, which are usually a vivid blue. These warts contain stinging cells with which it will attack other anemones of the same species if they start to invade its space. This anemone produces live young and will lift the new 'babies' from inside the body cavity on the end of one of its tentacles and place them around the adult.

2 ACTINOTHOE
Actinothoe sphryodeta
Most common around Gibraltar, Morocco and the eastern coast of southern Spain, this is a communal species, frequently found on shipwrecks and rocky surfaces. It has more than 100 tentacles, which are always white, with the disc ranging in colour from a fluorescent green to orange or brown, and 4cm (1.5in) across. It is able to fire sticky threads from the tentacles as a means of catching prey.

3 CLOAK ANEMONE
Adamsia carciniapados
A truly adaptive species, the Cloak Anemone wraps itself around the shell inhabited by the hermit crab *Pagarus prideaux* (page 146). The column of the anemone is pale fawn with garish magenta spots, and the oral disc contains more than 500 small tentacles. The disc is usually found on the bottom of the crab's shell and enjoys the hermit crab's messy leftovers. Its size depends on the shell it covers. When threatened, it secretes long, sticky, magenta threads.

4 TRUMPET ANEMONE
Aiptasia mutabilis
Living in holes or small crevices, this anemone has a slender column up to 17cm (7in) long, but this is rarely seen. The column flares out widely and has about 100 long, almost transparent, tentacles. The tentacles do not retract readily; rather, they have additional protection in the form of a coiled sticky thread which shoots out from the tentacle tip if threatened. The anemone has a bluish tint due to the presence of zooxanthellae.

1 BERRIED ANEMONE
Alicia mirabilis
Sometimes referred to as the 'Wonder Anemone' owing to its amazing stretching ability, *Alicia* is better recognised by the numerous small berry-like bumps that cover its body. These berries or tubercles are armed with stinging cells and can be quite harmful to human skin. During the day, it closes up and becomes a small inconspicuous lump only 10–15cm (4–6in) tall, with its tentacles retracted. At night it can stretch over 1m (3ft 3in) and its long tentacles trap prey in the current.

2 PLUMOSE ANEMONE
Metridium dianthus
One of the widest-ranging temperate waters anemones, it is found in all regions of the North Sea and extends through Iceland, Greenland and North America and into the Mediterranean. There are two different varieties, one large and one small, which are now widely recognised as separate subspecies. This is a rather beautiful anemone in shades of white or orange, often with the column a sickly green colour. When fully exposed and reaching as high as 45cm (1ft 6in), the column is smooth but for tiny holes where stinging threads can be ejected. There is a distinct collar underneath the crown of more than 200 small tentacles. These create a distinctly fluffy or plumed appearance, giving the anemone its name. This anemone will grow almost anywhere, from shipwrecks to piers and rocky reefs. It reproduces by pedal laceration.

3 SNAKELOCKS ANEMONE
Anemonia sulcata
One of the most common of the Mediterranean anemones, the Snakelocks Anemone is instantly recognisable by its 'gorgon-like' head of around 200 tentacles, which cannot be retracted as they are simply too numerous and so much larger than the body cavity. Generally grey or green in colour with purple-tipped tentacles, it is found on the lower shore where it is associated with a number of small crabs and gobies. It is up to 19cm (7.5in) across.

4 HERMIT ANEMONE
Calliactis parasitica
This large anemone, 5cm (2in) across, lives in association with a hermit crab. It has quite a stiff column, cream in colour with vertical brown stripes. The oral parts are ringed by more than 700 slender tentacles and the body is able to fire sticky threads or acontia when threatened. The hermit crab's shell may bear large numbers of these heavy anemones, making it rather difficult for the crab to manoeuvre – but it is well protected!

5 DAISY ANEMONE
Cereus pedunculatus
Not as common as other anemones, this species lives happily amongst small stones, rocks or even a rough gravel seabed. In various shades of brown, this anemone is tall and trumpet-shaped with a wide oral disc surrounded by 500–1,000 short tentacles. The column has some indistinct suckers to which it can attach bits of debris to create additional camouflage. It is 5cm (2in) across.

1 FIREWORKS ANEMONE
Cerianthus membranaceus
One of the largest Mediterranean species, *Cerianthus* prefers deep sheltered water or inside caverns, well away from strong sunlight. Secreting a parchment-like tube in the soft mud or sandy seabed, this anemone is distinguishable from other species by having two different types of tentacles. Inside the mouth are large numbers of small feeding tentacles, surrounded by very long, outer catching tentacles, which form an 'exploding firework' effect reaching over 1.5m (5ft). Sensitive to both pressure and light, this anemone is quite difficult to photograph as it is extremely retractile, able to withdraw its tentacles in a split second. It is very similar in character to *Pachycerianthus multiplicatus*, found in British waters.

2 GOLDEN ANEMONE
Condylactis aurantiaca
This large anemone lives in the soft sand in deep water, usually below 25m (80ft). The column is deep down in the sand, normally attached to a small stone. The oral disc is ringed with 96 fat tentacles with an overall diameter of around 20cm (8in). It is sometimes mistaken for the Snakelocks Anemone (see page 116) due to the coloration of its tentacles, which are usually greenish in colour and tipped in purple. Large numbers are found around the wreck of the *Rozi* in Malta.

3 CLUB-TIPPED ANEMONE
Tematactis crinoides
Similar in shape and character to the Dahlia Anemone, this species is commonly found around the Atlantic Islands of the Azores and Madeira, and also along the coast of Spain and Gibraltar.

4 DAHLIA ANEMONE
Cribinopsis crassa
Very similar to the *Urticina* species found in North Sea waters, *Cribinopsis* is quite rare and only found on its own. Preferring a rocky crevice or hole in which to hide its body column, its short, stubby tentacles are very retractile. It grows to around 12cm (4.75in) in diameter and comes in various shades of blue and green, owing to the presence of symbiotic zooxanthellae. This anemone is commonly associated with its symbiotic partner, the Amethyst Shrimp (see page 142).

5 JEWEL ANEMONE
Corynactis viridis
Not usually associated with the Mediterranean, this tiny anemone is surprisingly common on offshore sea mounts and vertical deep walls where it enjoys low sunlight and nutrient-rich water. With the polyps reaching only a maximum size of 10mm (0.5in), it is invariably still easy to spot owing to its iridescent bright colour. The short tentacles have knobbly tips. Able to reproduce asexually, many individuals of the same colour form are often found together, forming large sheets of colour.

Zoanthids

ZOANTHIDS ARE SHORT-BODIED colonial animals, very similar to anemones and other corals. They have no skeleton and spread by means of an extending or creeping stolon. The single mouth of each polyp is ringed by two rows of tentacles, containing photosynthesising cells known as zooxanthellae.

6 GOLDEN ZOANTHID
Parazoanthus axinellae
These are brilliant yellow- to orange-coloured polyps, which can grow as high as 2cm (0.75in) and have 24–36 tentacles in two circles. Preferring shaded areas, they are found under overhangs and at the entrances to caverns, and are known to colonise many other types of more sedentary forms of marine life such as sea squirts, pen shells and even sea fans.

Corals

THE CORALS FOUND in the Mediterranean are of the solitary polyp variety and, although one may find them in extensive colonies, the reef-building corals more commonly associated with tropical waters are not found here, owing to the lower temperature of the water. This may well change in the future with the influx of warm water carrying coral plankton through the Suez Canal from the Red Sea and greater Indian Ocean. Unlike anemones, corals are supported by a calcified skeleton.

1 STAR CORAL
Astroides calycularis
This brilliantly golden coral forms large sheets, which overgrow rocky surfaces. Reaching over 2cm (0.75in) when fully extended, its calcified skeletons are connected by a common coenosteum with the mantle or bodies of the animals seemingly connected because they are packed so tightly together.

2 CUP CORAL
Caryophyllia smithii
One of the few coral polyps that inhabits British waters, the cup coral is particularly common in the western Mediterranean. This familiar species is recognised by its knobbly-ended tentacles, spreading out from a deeply grooved calcified disc approximately 1.5cm (0.5in) in diameter. As in all cup coral species, the tentacles can be retracted, allowing the calcified exterior to provide full protection to the animal. When viewed under blue light, this cup coral fluoresces and shows up as a brilliant green!

3 MAT CORAL
Cladocora caespitosa
A migrant from the Atlantic, the brownish-green *Cladocora* is one of the few cup coral species that forms dense mats of individuals. Similar in structure to the reef-building corals of more tropical waters, this species has calcified 'heads' from 5–10mm (0.25–0.5in) in diameter. Colonies can grow over 1m (3ft 3in) across.

4 YELLOW CUP CORAL
Leptosammia pruvoti
This solitary coral polyp is distinguishable by its brilliant, almost fluorescent, yellow colour. Quite tall, the stem or corallum can be 2cm (0.75in) high. It prefers cave or cavern conditions as it is particularly light sensitive.

5 DARK SOLITARY CORAL
Phyllangia mouchezii
More notably found in the western Mediterranean, this is another species which has migrated into the Alboran Sea from the Atlantic. Solitary by nature, it has a pronounced brown calcified corallum reaching over 2cm (0.75in). The tentacles are almost transparent and knobbly.

6 LARGE SINGLE CORAL
Polycanthus muellerae
This is a large single polyp that grows in extensive, diffuse colonies. Preferring low light, it is usually found in caverns or under overhangs. Unlike other corals, it does not have any symbiotic relationship with microscopic algae for protection. It feeds on various planktonic creatures.

BRYOZOANS

BRYOZOANS ARE A PHYLUM of delicate, sessile, colonial animals, containing many different families with highly varied characteristics. They come into the category of false corals, due to their generally coral-like shape and structure.

1 LACE CORAL
Caberea boryi
Forming a delicate, creamy-white, lace basket, this small bryozoan has a calcified skeleton and grows to around 2.5cm (1in) in diameter. Found in depths of around 20–50m (66–165ft), it prefers a shaded location, but well aerated by water movement. It is commonly found in association with other species of bryozoans.

2 FALSE CORAL
Myriapora truncata
The most common of all the bryozoans found in the central Mediterranean, False Coral inhabits deep water and the entrances to caves and caverns where it enjoys the low light conditions. Tinged rosy red and orange, this distinctive bryozoan has widely spaced, short, rounded branches, which give the appearance of having been chopped off. Looking closely at the branches, you will notice minute pores. These contain the feeding parts of the organism, which are similar to coral polyps. Growing to a height of around 4cm (1.5in), it is found in depths from 3–90m (10–300ft), sometimes deeper. Often confused with Antler Coral.

3 DEER HORN CORAL
Pentapora fascialis
Forming rounded clumps around 15cm (6in) in height, this delicate marine organism generally grows upright in poorly illuminated areas such as caverns. Tinged from amber to burnt orange in colour, it has wide flattened branches that divide and fuse together, overlapping to create a domed mass, similar to a coral head. Particularly delicate, it is easily dislodged by careless divers.

4 ENCRUSTING HYDROID
Schizomavella mamillata
Very hard and coral-like to the touch, this rarely seen encrusting species is tinged salmon pink with raised creamy heads, which look like spots from a distance.

Preferring low light areas, it is quite tolerant of poor water conditions and is known to encrust the keels of ships.

5 FROND CORAL
Frondipora verrucosa
Very similar to *Caberea boryi* and often confused with this species, *Frondipora* has less of an obvious structure and does not form the basket-like shape. Rather, the fronds branch at different intervals and are thicker in structure. This species grows to a maximum size of 2.5cm (1in) and also enjoys fairly deep water.

6 BRANCHING HORN WRACK
Securiflustra securifrons
Found in the North Sea, Atlantic and western Mediterranean, this pale creamy colonial bryozoan is stiff and calcified, but is not rigid like some of the other species, behaving more like a type of alga than a colony of animals. It can form quite dense clumps up to 10cm (4in) high, and prefers a rocky substrate in well-aerated water.

7 NEPTUNE'S LACE
Sertella septentrionalis
Easily distinguished by its delicate, pinkish, lacy whorls, Neptune's Lace forms a cup-like shape with fluted edges. Over 5cm (2in) in height, it enjoys a cavern environment with low light conditions and is commonly associated with several other species of bryozoans, which enjoy the same habitat.

8 ANTLER CORAL
Smittina cervicornis
Antler Coral is often mistaken for False Coral as they often intermingle on the same cliff wall. When using torch light, you can quickly see that not only is there a colour difference, but the structure is also different. Antler Coral is cream-coloured and its branching segments are much more tightly packed together. It grows up to 10cm (4in) long.

MOLLUSCS

THE MOLLUSCS FORM the largest and most diverse phylum of marine animals on the planet, and are among the most familiar groups of invertebrates. The family includes nudibranchs or sea slugs, chitons, snails, bivalves, octopus and squid. There are some 75,000 species in eight different families found in the world's oceans. Most species have common characteristics such as an external shell, a soft body with a muscular foot, and a rasping, tongue-like radula for feeding, with gills of some kind for respiration. All lay eggs, and, of the eight groups, three are fairly well represented around these shores.

Cephalopods

THE CEPHALOPODS ARE perhaps the most advanced of the molluscs, and octopus are common throughout the Mediterranean. Extremely intelligent and adaptable, they are free-swimming, fast-moving predators with well-developed eyes and sensory systems. All of the cephalopods are able to change their shape and coloration to suit their environment or to display to a mate or threaten a would-be predator. They are also considered prized catches and feature in most coastal seafood restaurants.

1 COMMON OCTOPUS
Octopus vulgaris
Common indeed, this species of octopus is very wide-ranging and is found and commercially fished all around the Mediterranean. More than 20,000 tonnes are taken annually off the North African coast alone. With a total body length of around 1m (3ft 3in) and maximum arm spread of 3m (10ft) in diameter, the body is stout and warty in appearance with two rows of suckers on the arms. It preys on crabs and lobster, and will also eat shellfish. Its den is usually discovered due to the amount of shell debris found outside.

2 MUSKY OCTOPUS
Eledone moschata
Usually mistaken for the Common Octopus, this species is fairly common in shallow waters around the western Mediterranean. With a body length of 15cm (6in) and arms to around 40cm (1ft 4in), it is predominantly brown with darker brown spots arranged irregularly over the body and webs. The animal is distinguished by its single row of suckers. The common name comes from the musky smell it exudes when taken out of the water. The female of the species lays around 500 eggs, each about 10mm (0.5in) long.

3 COMMON CUTTLEFISH
Sepia officinalis
This species grows to over 45cm (1ft 6 in) long and has an internal cuttlebone, which is a chambered, gas-filled shell used for buoyancy control. Possessing eight arms and two longer feeding tentacles, the adult is recognised by the white bands across the body and small white spots on the upper surface of the fins. The arms have four rows of suckers. It actively stalks its prey around *Posidonia* beds and inshore reefs. Like all cuttlefish species, it is characterised by its ability to change colour and body texture at will.

4 WHITE-SPOTTED OCTOPUS
Octopus macropus
The White-spotted Octopus is more commonly seen at night, when it is an active hunter. It appears to be exclusively nocturnal and there are virtually no daytime records of the species. The body grows to around 15cm (6in) and the arms are over 1m (3ft 3in) long. It is recognised by the pairs of white spots down the long arms and a reddish-brown body.

1 HOODED CUTTLEFISH

Sepia prashadi

Often pale-coloured, this species can change its body coloration and shape at will and blends in with most of its surroundings. Only growing to around 30cm (12in), it is an active hunter at night. It prefers shallow lagoons, sheltered bays and around coral rubble. The Hooded Cuttlefish is an immigrant from the Red Sea through the Suez Canal.

2 CALAMARI SQUID

Loligo vulgaris

Growing to over 75cm (2ft 6in) long, this is the most common squid found in the Mediterranean. It inhabits the water column to all depths, but comes into shallow coastal waters to spawn. The female lays around 20,000 eggs in pale white or grey gelatinous strings attached to the sea floor or hard objects. This squid is caught commercially by a variety of trawl nets and with night lights and jigs.

Gastropods (Snails)

GASTROPODS FORM THE largest subcategory of molluscs, and their body forms are widely varied. The group includes all univalved shellfish, such as limpets, periwinkles and snails. They usually have a single-piece shell and live mostly on the shore or sea bottom. This group also includes the opisthobranchs, which have no shell at all.

3 GREEN CHITON

Chiton olivaceus

This ancient species of animal is particularly distinctive. It has eight hinged overlapping plates, which allow the animal to fold itself up. Oval-shaped and tinged with various shades of green, it is able to clamp onto rocks with the aid of a strong muscular foot. Common on the shoreline and surf zone, it enjoys aerated water and grows to a maximum size of around 5cm (2in).

4 WHELK

Buccinulum corneum

This gastropod has a thick, spindle-shaped shell with an oval opening and an upturned tip where its eye stalks protrude. Usually pink and beige in colour, it grows to around 6cm (2.5in) and enjoys grazing amongst *Posidonia* and algae-encrusted rock surfaces.

5 RAPA WHELK

Rapana venosa

First discovered in the Black Sea in 1947, this alien invader was thought to have arrived in a ship's ballast from Asia. It soon took over the region and was considered an environmental disaster. Quickly taking hold of its new environment, the mollusc was so successful that not only is it now harvested commercially, it is also exported from the Black Sea back to the Far East as a highly successful commercial product.

6 TRUNCATED WHELK

Trunculariopis trunculus

Growing to around 8.5cm (3.5in), this whelk has a very rough exterior shell. It grazes on various algae and, although it is often seen alone, it can form large breeding groups and is known to mass spawn. With only a few concentric whorls of the shell, the anterior is markedly upturned and protects the eye stalks.

7 MEDITERRANEAN CERITH

Cerithium vulgatum

This tower shell grows to 6.5cm (2.5in) and has a thick and broad structure with prominent ridges at the growth lines. It has quite a large aperture and thick, flared lips. It is common in soft sand and shell debris areas where it filter feeds. Its shell is a common home for hermit crabs and you can often see it in large numbers on a dead fish or crab.

8 ROCK CERITH

Cerithium rupestre

This species is also a pointed conical shell, but the whorls are more rounded and not as indented as *Cerithium vulgatum*. Very common, it prefers a rocky shoreline and also grows to around 5cm (2in).

1 GIANT TUN SHELL

Dolium galea

Spends much of its time buried in fine sand or mud, and comes out at night when it is an active feeder on sea cucumbers. This species can grow as large as 30cm (12in), but has no protective operculum, making it easy prey for octopus, which can sometimes be found hiding in its empty shells. This specimen was photographed at Mgarr Ix Xini on Gozo.

2 WOODY HORSE CONCH

Fasciolaria lignaria

This small Murex snail was first described by Linnaeus in 1758 and is common throughout the Mediterranean. With around seven whorls and raised bumps on the outside rim of these whorls, it is usually quite brightly coloured. It grows to around 25 mm (1 in).

3 PURPLE DYE MUREX

Bolinus brandaris

An ancient species, this predatory sea snail is known from fossil records from the Pliocene (3.6–2.6 million years ago). Not only is this snail a highly prized food, its mucus was the source of the Imperial 'Tyrian purple' dye.

4 PELICAN'S FOOT SHELL

Aporrhais pespelicani

This shell is quite wide at around 2.4cm (1in) and is generally spindle-shaped with a very developed structure. It has around 10 whorls. Pale cream in colour and preferring muddy or fine sand sediments, it has a long, irregularly formed aperture with distinct splits. The shell forms four or five blunt tips and looks similar to the splayed foot of a pelican or other seabird (hence the name). It has a long drawn-out mouth with two rows of radula (or rasping plates) with seven teeth on each. It has a short siphon.

5 NECKLACE SHELL

Natica alderi

This snail is only around 15mm (0.5in) in height and hides under the sand or mud during daylight hours. An active hunter at night, it leaves very distinctive egg whorls on the seabed.

6 ROUGH SNAIL

Bolma rugosa

This predatory snail can grow over 50mm (2in) and has six or seven whorls, usually brownish grey in colour and rather plain. It has an incredibly colourful protective operculum, which is often used in jewellery manufacture, and the interior of the shell is pearlescent.

7 MOTTLED TRITON

Pisania maculosa

This is a small sea snail, growing to around 25mm (1in). It vaguely resembles a tiny Triton Trumpet. In some areas, it is known to eat barnacles and other small molluscs and crustaceans.

8 TRITON TRUMPET

Charonia tritonis

The largest of the trumpet shells and quite rare in the Mediterranean, this species feeds on starfish and is an active predator. It can grow to over 40cm (1ft 4in) and is constructed of irregular spires, often overgrown by calcareous algae. The foot is beige, spotted with brown, and the eye stalks have two dark bands.

1 MEDITERRANEAN COWRIE

Cyprea lurida

Fairly common on night dives, this species of cowrie grows to around 3cm (1.25in) and is found in central and eastern areas of the Mediterranean. It has brownish, broad bands across its back and at both ends. The mantle of the animal is also brown with spiky protuberances and when fully extended is difficult to spot.

2 SPOTTED COWRIE

Cyprea spurca

This colourful cowrie is relatively rarely seen, as it is a night-time predator, preferring to hide well away in a rocky crevice during the day. More likely to be found in a cave situation, this cowrie is tinged orange and rusty brown and eats various types of sponge and bryozoan. The shell is kept smooth and shiny by the animal's mantle skirt, which completely covers the shell whilst moving.

3 THREE SPOT or EUROPEAN COWRIE

Trivia monacha

This mollusc has an egg-shaped glossy shell featuring many transverse ridges, with a long, narrow aperture on the underside. It grows to around 15mm (0.6in) and feeds on soft bryozoans and hydroids. The upper surface of the shell is usually a reddish-brown colour, and has three characteristic spots that allow the species to be identified easily. The head, tentacles, foot and body are brightly coloured; they may be red, yellow, green, brown or orange.

4 MEDITERRANEAN CONE SHELL

Conus mediterraneus

Distinctively shaped, this cone shell can grow to 7cm (2.75in). The shell is typically broadest at the abapical end of the slit-like aperture, which tapers towards the open siphon canal. It has a short spire with only three or four whorls and the entire shell is smooth and shiny with mottled brown striations.

5 TOP SHELL

Calliostoma conulum

A splendid smooth-sided conical shell with around 10 straight whorls. Colour is variable from a light tan to almost a dark purple, and it may or may not have other markings on the shell, but it is the obvious shape which distinguishes it from any other species. It can grow to around 35mm (1.5in).

6 ABALONE

Haliotis lamellosa

Confined to the Mediterranean, the Abalone grows to 7cm (2.75in) and prefers a shaded position, more often than not hiding under rocks during the day. It is recognised by its low spire and limpet shape, and the outer surface has a keel of small round holes. The fleshy foot is not only able to attach firmly to a rock, but also allows for easy propulsion. It feeds on red algae. The inside of the shell is of iridescent mother of pearl.

7 TUBE SNAIL

Lementina arenoria

This is a curious mollusc from a diverse family, comprising nine species that are easily confused with each other. Resembling a tube worm, the Tube Snail has a wide, obvious shell and can grow up to approximately 1.5cm (0.5in) in diameter. The animal has no operculum and the mouth of the tube shows the reddish-brown snail streaked with a golden-yellow pattern. It lives inside this tube all the time and fishes for prey by extending a mucous net, capturing plankton and transporting it to the gut by ciliary currents produced by its gills.

Sea Hares

SEA HARES ARE shell-less gastropods, characterised by rather fleshy bodies, which are swollen towards the rear. The head is split into pairs of lobed oral and sensory tentacles. Sea hares have the ability to release clouds of purplish ink, as well as milky white, sticky threads as a means of defence.

1 GIANT SEA HARE
Aplysia fasciata
Growing to over 40cm (1ft 4in) and with a weight of nearly 2kg (4.5lb), the Giant Sea Hare is dark peaty brown in colour with a mauve-tinged edge to the skirt. It feeds on a large variety of seaweeds and is usually found in pairs.

2 SMALL SEA HARE
Aplysia punctata
Growing to only 20cm (8in), half the size of its cousin, the Small Sea Hare is also a much lighter brown in colour, although this colour varies depending on the diet. It is delicately marked with lighter spots over a mottled or blotched body.

Pleurobranchs

PLEUROBRANCHS STILL HAVE the vestiges of a shell, which may be either internal or external. They have acid-secreting glands for defence and most species are carnivorous.

3 UMBRELLA SNAIL
Umbraculum mediterraneum
This is a very distinctive oval slug, 4cm (1.5in) long. It looks as though it is wearing an algae-covered limpet as a hat and, owing to the hundreds of lobed portions to the body, is sometimes mistaken for a shell with its egg cases. It has two short rhinophores, or sensory tentacles, at the head and enjoys light and well-aerated water.

Bivalves

4 QUEEN SCALLOP
Aequipecten opercularis
This small scallop is highly prized in the fish market and is commercially harvested throughout the Mediterranean and Atlantic. It grows to around 9cm (3.5in) in diameter and has 20 sculpted, bold ridges on each side. It lives on mud and soft, sandy seabeds. It usually becomes overgrown by a variety of sponges and it is extremely mobile when danger threatens. With the aid of its muscular joint, it opens and closes the two halves of the shell rapidly, thus providing a jet-propelled escape.

5 THORNY OYSTER or ROCK SCALLOP
Spondylus spp.
This species is probably *Spondylus gaederopus* but, as always with these bivalves, it is difficult to identify, as the covering of distinctive raised spines on the shell is covered by the sponge *Crambe crambe*. The bottom half of the shell is firmly adhered to the rocky surface and reaches 12cm (5in) in length. Preferring caves and low light conditions, this scallop has a distinctive double row of eyes around the body cavity, and a pale cream inner muscle. Highly prized both for its meat and its pearl-producing capabilities, the species is in decline throughout the Mediterranean.

6 COMMON MUSSEL
Mytilus edulis
Widespread in shallow water, particularly around the surf zone, mussels are a highly prized addition to the Mediterranean seafood diet. Commercially farmed in a number of areas, wild mussels have dark bluish-purple shells and can attach themselves together in large groups by means of a renewable byssus – a stiff, secreted thread. Mussels grow up to 5cm (2in) long and there are some 36 species found in European seas.

1 WINGED OYSTER
Pteria hirundo
This very distinctive oyster has equal-sized valves to the shell and is usually found attached to Gorgonian Sea Fans, where it exploits the strong current to filter-feed on passing plankton. Utilising byssus threads to attach itself to the sea fan, this shell grows to 6cm (2.5in) in diameter. It is vaguely oval but with a distinct thin ear or wing.

2 SPINY FILE SHELL
Lima lima
This file clam always lives in a rocky crevice and is usually seen when it extends its white, sticky, feeding tentacles from the mouth of the shell. It soon grows too large to be able to move from its home and will eventually die in the same location. Those found in the open are able to swim in the same way as a scallop does, by expanding and contracting a strong muscle, opening and closing the shell rapidly, providing propulsion.

3 NOBLE PEN SHELL
Pinna nobilis
Resembling a partly folded fan, this giant mussel is highly sought after as a souvenir and is a protected species in the Mediterranean. In special circumstances, the shell can attain a height of over 1m (3ft 3in). It prefers to live amongst *Posidonia*. The outer shell is quite rough and makes an ideal home for many different species of algae and sponge; it also harbours a few commensal crabs and shrimps within its body cavity. In the nineteenth century, the byssus threads were collected, dried and then woven into gloves, shawls and stockings, similar in texture to silk. These products were greatly sought after by the upper classes.

4 ROUGH PEN SHELL
Pinna rudis
Much smaller than its cousin, this pen shell grows to around 10cm (4in) across and has a much flatter anterior lip. This fan mussel prefers very low light conditions and a hard rocky substrate where it grows from a crevice. Also found on rough gravel sea floors at the entrance to caverns, this rare species has delicate raised whorls along the outside of the almost translucent shell.

5 GOLDEN CARPET SHELL
Paphia aurea
This small bivalve is found in large numbers in rough sand and shell gravel and is harvested mercilessly. Growing to 3cm (1.25in) across and roughly triangular in shape, the concentric structure has obvious growth rings and is overlaid with an intricate zigzag pattern.

6 SMOOTH CARPET SHELL
Pitar rudis
Extensively collected for public consumption, the Smooth Carpet Shell, up to 2.5cm (1in) long, has a shiny outer surface in shades of beige and golden-brown, with darker concentric circles of colour. It prefers a soft sand or muddy seabed, where it can occur in vast numbers.

7 WARTY VENUS
Venus verrucosa
This is a stout bivalve with a shell of equal halves formed of more than 20 concentric growth rings. Growing to around 6cm (2.5in) in diameter, it is yellowish white or orange in colour and lives partly buried in rough sand and shell gravel. This species is highly prized by octopus, which often use the shells to seal the entrance to their dens.

Nudibranchs

THE NAME 'NUDIBRANCH' means 'naked gills' and these are snails which have lost their shells – hence often being known as sea slugs. They are highly diverse, but shared characteristics include obvious rhinophores at the head in most species, and a set of gills at the rear of the body. All are carnivorous.

1 WHITE-TIPPED NUDIBRANCH
Edmundsella pedata
This species grows up to 5cm (2in) long and has seven paired clusters of cerata (tentacles along the back of the body), all of which are tipped with white. The body is a uniform purple and may have some light spots along its back. It feeds on various hydroids, including *Eudendrium*. Not to be confused with the Purple Nudibranch.

2 MIGRATING AEOLID
Cratena peregrina
Quite common in central and northern regions, where it can occur in quite large numbers, depending on the time of year. It has six clusters of cerata, shaded brown with a purple tip. The oral tentacles are tinged blue to white and there is a distinct orange flash on the head between the rhinophores. It grows up to 5cm (2in) long.

3 TRI-COLOUR DORIS
Hypselodoris tricolor
This species grows to only 2cm (0.75in) and is a brilliant blue in colour with equal horizontal yellow lines and another yellow line completely encircling the body. The rhinophores are dark blue, as are the gills.

4 PURPLE NUDIBRANCH
Flabellina affinis
A garish violet colour, it is also deeply tinged with red. Unlike *Edmundsella pedata*, it does not have white tips to its cerata. It feeds on the polyps of *Eudendrium* hydrozoa. Growing up to 5cm (2in) long, it has six to nine paired clusters of cerata and its rhinophores are the same as the body colour.

5 LADY GODIVA
Dondice banyulensis
Quite common throughout the region, this aeolid has a beige body with white lines down its back and outside of the skirt. The oral tentacles are tinged with blue and the seven clusters of cerata along the back are tinged with fawn and brown with golden tips. It grows up to 5cm (2in) long.

6 SEA LEMON
Dendrodoris limbata
Typically oval in shape and lemon-yellow to brown in colour, it has a blotched body with a yellow frill. This species is quite small, only growing to 2.5cm (1in), and inhabits deeper water where it feeds on a variety of hydroids, bryozoans and even sponges.

7 RED ROSTANGA
Rostanga rubra
A small but very obvious sea slug due to its orange coloration. Its rhinophores are a pale cream and quite stunted and the main part of the body is covered in minute tubercles and defensive spicules. This species is wide-ranging and feeds primarily on sponges. It grows to around 15mm (0.5in).

8 YELLOW MARGIN DORID
Diaphorodoris luteocincta
Growing to around 2.5cm (1in), it is easily distinguishable from many others by the distinctive markings along its back and margins. The centre of the back is brick red, fading to white, and has a brilliant yellow fringe, also lined in white. The back is also covered in large, white, spiky tubercles. It has two long, white, spiral rhinophores. It eats bryozoans and is only found subtidally, usually below 6m (20ft).

9 YELLOW-TIPPED NUDIBRANCH
Limacea clavigera
Easily recognised, yet small, it grows to around 2cm (0.75in). It is white in colour, has a stout, fringed mantle and all of its rhinophores, tentacle and various tubercules are tipped in brilliant yellow. It has a number of extended tentacles at the front of the head, which also extends below the margin. It is common throughout the North Sea, Atlantic and Mediterranean. It eats bryozoans.

1 WHITE LINED DORID
Doriopsilla pelseneri
Another distinctive species, this dorid is golden-yellow and grows to 3cm (1.25in) long. The oval body is covered in a tracery of fine white lines. Both the rhinophores and gills are the same colour as the body and are retractile.

2 ELEGANT SEA SLUG
Hypselodoris elegans
This is one of the largest nudibranchs to be found in the Mediterranean, growing to over 15cm (6in). It comes in a variety of body colours, all of which are rather bright and garish, and include brilliant yellow, violet and even green. The gill tuft is very prominent, almost like a flower, and the body mass has a frilly edge. It prefers fairly deep water, feeding on encrusting hydroids and coral polyps. There would appear to be some confusion over this species of Hypselodoris as it is known by various other names, depending on the colour pattern. Scientists are still not convinced that these other colour forms are separate subspecies, rather than just colour changes due to different diets. Names that you may come across in various regional guide books are *H. picta*, *H. webbi* and *H. valenciennesi*.

3 SPOTTED POLYCERA
Polycera capita
Once thought to be the same species as *P. quadrilineata*, this nudibranch has recently been reclassified. It resembles *P. quadrilineata* in every way except that it is more colourful in appearance, being covered in intricate colourful spots and lines.

4 CEUTA SEA SLUG
Tambja ceuta
Largely confined to the Strait of Gibraltar and the Moroccan coast, this species is a dull greyish green or blue with darker blue or violet lines and pale yellow along the body. It grows to over 4cm (1.25in) long.

5 FOUR-LINED NUDIBRANCH
Polycera quadrilineata
A common sea slug in the western Mediterranean, this species is an Atlantic immigrant and is common throughout the Bay of Biscay and up to Scandinavia. It has a distinctive white body with yellow stripes around the frame. It has four prominent yellow-tipped oral tentacles and grows to 3.5cm (1.5in), and feeds on a variety of bryozoans and hydroids.

6 ELYSIA
Elysia timida
This colourful and delicate opisthobranch grows around 2.5cm (1in) in length and has rudimentary eyes situated behind the rhinophores. It tends to be shades of green in colour, owing to the absorption of chlorophyll from the algae that it eats. This common species has been observed feeding on the invasive *Caulerpa taxifolia*.

7 LEATHERY DORIS
Platydoris argo
Burnt orange or brown in colour and roughly oval in shape, this nudibranch's outer skirt is deeply ruffled, giving it a rather indistinct outline. This curious doris is always found in association with a partner or partners, usually moving along in pairs, just slightly overlapping. It can grow to 10cm (4in) long and its rhinophores and gills are deeply set back from its frame and are totally retractile.

8 SPOTTED DORIS
Discodoris atromaculata
Particularly distinctive in appearance, with a light-coloured body and dark brown, irregular spots, this species usually grows to around 12cm (4.75in) in length, but may be considerably smaller. The eight gills and rhinophores are white. This species feeds almost exclusively on the sponge *Petrosia ficiformis* and can be found in open, well-lit areas as well as deep caverns and caves.

9 CRYSTAL TIPS
Antiopella cristata
This species has an oval, flattened shape and is fringed all around the 7.5cm (3in) body by golden cerata or tentacles, each tipped with white and iridescent blue. The rhinophores are white and non-retractile. This species feeds on bryozoans. Widely distributed throughout the Mediterranean, Canary Islands, Bay of Biscay and the British Isles, it is an active night-time predator. It can be found on many different substrates, preferring clean, shallow and sheltered waters.

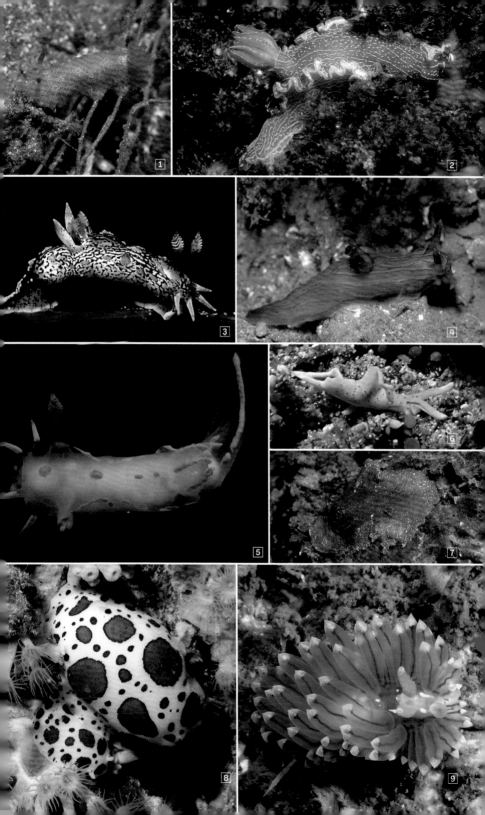

CRUSTACEANS

THIS LARGE GROUP of arthropods, related to insects, has more than 30,000 species worldwide. Crustaceans can be found in every marine habitat. In general they have segmented bodies with a head, thorax and abdomen, and are heavily protected by calcareous outer body armour. Most divers see the obvious decapod (ten-legged) species. The legs, depending on the species, are adapted for walking, feeding, swimming, defence, food capture, respiration and even the carrying of eggs.

Sea Spiders

1 SEA SPIDER
Endeis spinosa
This sea spider is regarded as very rare. It is quite small at only 10–15mm (0.5in) long, and is difficult to find. The body is smooth and pale cream to light brown in colour, and is rigidly segmented. The legs are of equal length and may well have bits of detritus attached to them. Widely distributed, it is commonly found at the base of sea fans and around hydroids and various calcareous algae, where it is able to hide its vulnerable body should danger threaten.

Barnacles

2 STAR BARNACLE
Chthalamus stellatus
When still in their juvenile, planktonic form, barnacles display their feeding arm. It looks like a catching mitt, and that is what it is. Distant relatives of the crabs, barnacles settle onto a suitable habitat in the shallow water surf zone, where they develop a protective shell of calcified rigid overlapping plates and become sessile. Only a few millimetres across, they are packed closely together. They are able to seal themselves off in bad conditions or exceptionally low tides, and are able to survive for a short while out of the water.

3 MEDITERRANEAN BARNACLE
Balanus perforatus
This creamy-purplish barnacle often grows alongside mussels and limpets and is cone-shaped, attached to hard substrate and often exposed at low tide. Usually forming small clumps, it competes for space.

4 GOOSE BARNACLE
Lepas anserifera
The Goose Barnacle has a similar method for catching prey as the Star Barnacle, but although this species is also sedentary in nature, it attaches itself onto floating bits of debris, ships' keels and piers, and so may be carried long distances; it is tolerant of poor water conditions. The Goose Barnacle is around 4cm (1.5in) long and has a long, leathery foot and a strong, calcified, protective head where the catching mitt is kept.

Shrimps

① OPOSSUM SHRIMP
Paramysis helleri
Found throughout the Mediterranean and Black Sea, the Opossum Shrimp occurs in large numbers and is an important food source for small fish. Growing to only 1.1cm (about 0.5in) in length, it lives close to the sea floor or rocky crevices where it can quickly hide from predators. Vast numbers of Opossum Shrimps can be found in the cave systems along the coast of northern Spain, the Balearics, Corsica and Sardinia during the summer months. Deep inside the caves, they are preyed upon by slipper lobsters, Unicorn Shrimps, Common Shrimps and even anthias, which will venture away from the light of day after this veritable feast.

② ELEGANT SHRIMP
Palaemon elegans
This is one of the more common shrimps found around the coastline and is normally associated with a rocky subsurface. Quite a stout species, growing to around 6cm (2.5in), it has strong, dark reddish-brown stripes on a fairly transparent body. Generally found in association with Dead Men's Fingers and preferring low light conditions, this shrimp is an active forager by dusk and at night. It is an important human food source and is harvested commercially in a number of Mediterranean countries that have vast sandbanks and shallow sheltered bays, such as the north African coast.

③ COMMON SHRIMP
Palaemon serratus
Unlike the Elegant Shrimp, this species will live in pitch-black conditions and is commonly found deep inside caves. It is also much larger at 11cm (4.5in). Although this species of shrimp enjoys a communal lifestyle with large numbers of its own species or alongside Unicorn Shrimps, more often than not it is found on its own, wandering around deeper caverns in search of Opossum Shrimps or edible detritus.

④ AMETHYST SHRIMP
Periclimenes amethysteus
One of the most colourful of all the Mediterranean shrimps, the Amethyst Shrimp has a symbiotic relationship with a few anemones, but principally *Cribinopsis crassa* (see page 118). Stout of body, up to 2.5cm (1in) long and nearly transparent, it has pinkish, well-defined markings with long, beige pincers and blue-striped legs. Seemingly rather docile when near its anemone protection, it is able to dodge very rapidly out of the way should danger threaten. The species can also be found around *Posidonia* seagrass beds, where its bright colours are a rare delight for divers.

⑤ WHITE STRIPE SHRIMP
Hippolyte inermis
Often also referred to as the Sea Grass Shrimp, this species is widely distributed throughout the eastern Atlantic, North Sea and Mediterranean. Mainly green in colour, it grows to around 4cm (1.5in).

⑥ UNICORN SHRIMP
Plesionica narval
More common in the central and western Mediterranean, the Unicorn Shrimp is found deep in caverns and is active at night. It has a distinctive clear body with red and gold lateral lines and quite often has blue eggs attached to the underside. It can grow to 10cm (4in). Like many other shrimp species, it starts life as a male and changes to female with maturity.

⑦ CLEANER SHRIMP
Stenopus spinosus
The Cleaner Shrimp is very timid and only rarely seen, and then only at dusk or during the night. It is golden-orange or rust-brown with long, white feelers with which it communicates with the various fish species that it cleans of parasites. Performing an important role in the health of the reef, this small shrimp is closely related to the Banded Coral Shrimp of more tropical waters. It is 10cm (4in) long.

Lobsters

1 STRIDENT SQUAT LOBSTER
Galathea strigosa
There are a few species of squat lobster in these waters, all migrants from the Atlantic. Principally found in the western Mediterranean, they have become quite adaptable to warmer waters. The most colourful of all is the Strident Squat Lobster with its brilliant orange-red body armour and vibrant, iridescent blue stripes across its head and around the eyes. It grows up to a maximum length of around 10cm (4in) and prefers low-light, shaded conditions.

2 BROWN SQUAT LOBSTER
Galathea squamifera
A small squat lobster growing to around 5cm (2in). It is a uniform brown to orange in colour and is found in small crevices, caverns and amongst stones on the seabed.

3 LONG-CLAWED SQUAT LOBSTER
Munida rugosa
Often referred to as 'langoustine' by the French, the Long-clawed Squat Lobster enjoys a deep-water habitat where it lives under stones and rocks. Often associated with brittle starfish beds and ascidians, it is instantly recognisable by its bright orange, fairly slim body, which is around 30cm (12in) long, and its chelipeds or pincer arms that are twice the body length.

4 NORWAY LOBSTER, SCAMPI or LANGOUSTINE
Nephrops norvegicus
This small lobster lives in large interlaced burrows under soft sand and mud. Pale orange to light tan in colour with transverse stronger coloration over the body segments, it typically grows up to 24cm (9in). However, some super-sized animals have been caught in traps and are over 38cm (15in), rivalling the size of the common lobster. This striking lobster has two large pincers and a secondary set of smaller feeding pincers. It has quite long antennae and jet black eyes, preferring deep-water dark habitats. It has a symbiotic relationship with Frei's Goby, and is commonly found in the Adriatic Sea.

5 SMALL LOCUST LOBSTER
Scyllaris arctus
Extremely light-sensitive, this very shy, small slipper lobster lives in deep caves with very little light. Growing to only 12cm (4.75in) in length, it has plate-like antennae, which have rounded, lobed edges. It is mottled brown with splashes of red in the joints of the abdomen, and a red ring around the slightly stalked eye. It has a very broad tail to allow for rapid movement and its eight obviously striped legs are used to grip onto the undersides of cavern ceilings and walls. When approached, it literally hops in reverse very quickly out of danger.

6 SHOVEL-NOSED or PADDLE-NOSED LOBSTER
Scyllarides latus
Robust, with no obvious pincer-like claws. It has overlapping body plates which protect its mouth parts. Usually found near seagrass beds, it is more active in the evening and at night. Non-swimming, it walks over the seabed, where it feeds on detritus with the use of a pair of adapted antennae.

7 SPINY LOBSTER
Palinurus elephas
This is a large spiny lobster, coloured in shades of orange and red, with a spiny carapace and abdomen. Growing over 50cm (1ft 8in) in length, it lives in caverns and rocky crevices. It is more active at night. It guards its lair, standing at the entrance on stout, long, striped legs with its huge, sensory antenna protruding into open water. Very timid in character, it is commercially fished using traditional lobster pots.

8 COMMON LOBSTER
Homarus gammarus
This lobster is more associated with Atlantic and North Sea waters, but it is surprisingly common in the Mediterranean. Preferring to live under large boulders or in deep rocky crevices, the Common Lobster is a nocturnal feeder. It is bluish in colour with stout pincers, one of which is modified for gripping, the other modified for cutting.

Crabs

Hermit Crabs

1 SEDENTARY HERMIT CRAB
Calcinus tubularis
Curiously, it is only the female of this species that is sedentary in nature, preferring to live in old worm snail holes. The male is much more adventurous. Slim-bodied, it is commonly found inhabiting empty cowrie shells. It grows up to 10mm (0.5in), has long stalked eyes, is reddish brown with pale striped legs and has pincer arms with red spots.

2 RED HERMIT CRAB
Dardanus calidus
One of the larger hermit crabs at over 2.5cm (1in), it is instantly recognisable by its bright red coloration, green eyes set out on long stalks, and orange antennae. It is often associated with the parasitic anemone *Calliactis parasitica*, which it lugs around on the back of its heavy shell home.

3 STRIPED HERMIT CRAB
Pagarus anachoretus
This species is fairly recognisable by its overall brownish coloration and white to bluish stripes and bands over all of its legs and pincer arms. It has a few long hairs on its legs. Its antennae are also long and banded with white. Its body length is 2.5cm (1in).

4 HAIRY HERMIT CRAB
Pagarus cuanensis
This is quite a small and distinctive species, growing to 1.5cm (0.5in), and is characterised by its extremely hairy lower legs and pincers, which always appear to be clogged up with bits of detritus. It is brown with long, stalked, pale blue eyes and brown-and-white striped antennae.

5 PRIDEAUX'S HERMIT CRAB
Pagarus prideaux
This species is mainly associated with the commensal anemone *Adamsia carciniapados* (see page 114), which completely envelops its carrier shell. The crab is speckled with white and tinged with mauve and is otherwise light tan with pinkish-striped legs and striped eye stalks. It grows up to 3cm (1.25in) long. The females lay eggs throughout the summer season. They are bright gold and are kept in the shelly home until the larvae are freed by the parent crab, endangering its softer body parts as they climb out of the anemone-covered shell.

Other Crabs

6 MASKED CRAB
Corystes cassivelaunus
This soft sand- and mud-living crab grows to around 7.5cm (3in) in length and has two long antennae, which form a breathing tube when the crab is hidden under the surface. Coloured a uniform light tan or pale cream, the carapace is roughly triangular in shape. The pincer arms or chelipeds are very long in the male and short and stout in the female.

7 SPONGE CRAB
Dromia personata
Muddy brown and covered in fine hairs, the Sponge Crab lives in low light areas, generally in caverns or caves, where it sometimes places a sponge on its back for additional camouflage. Growing over 7cm (2.75in), it is quite a stout crab with strong limbs and pincers, which have pinkish tips.

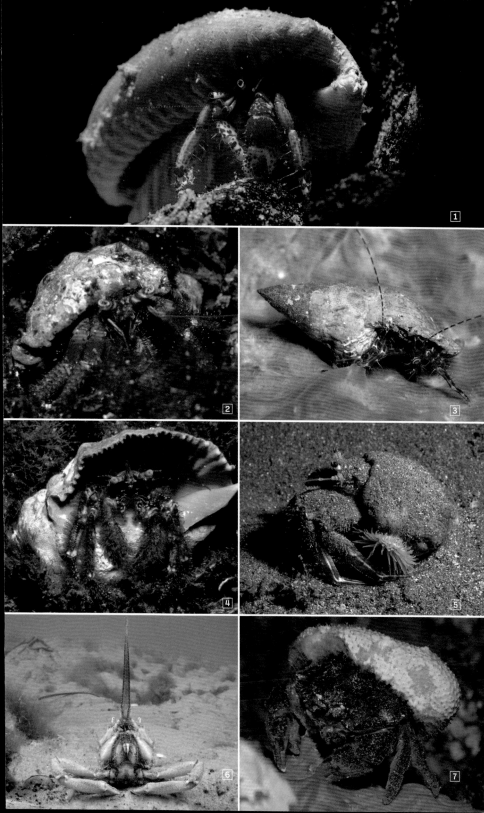

1 ANEMONE SPIDER CRAB
Inachus phalangium
Commonly associated with the Snakelocks
Anemone (see page 116), this spider crab
enjoys the protection of the anemone, being
seemingly immune to its stinging tentacles.
It is only the female of the species that lives
in the anemone. It is brownish and often
covered in fine hairs and bits of algae and
sponge, to help in camouflage. It only leaves
the anemone at night to feed but stays close
by. It grows to 5cm (2in) in length.

2 ARROW CRAB
Stenorhynchus lanceolatus
Almost identical to its Caribbean
counterparts, this crab is more commonly
found around the Azores, Madeira and the
Canary Islands, but it is gradually invading
the western Mediterranean and can be
found around Gibraltar and the east coast
of Spain. It stands quite upright and has two
tall spiky antennae and very mobile pincers.

3 SPINY SPIDER CRAB
Herbstia condyliata
This small species grows to only 3cm
(1.25in) and is widely distributed. An active
scavenger, it only comes out at night and
prefers underhanging rock faces dotted with
small holes. The legs are striped and it is
otherwise a reddish-brown colour.

4 RED SPIDER CRAB
Lissa chiragra
Rarely seen, and then only at night, the
Red Spider Crab grows to only 5cm (2in)
and is quite happy on rocky substrates or
amidst *Posidonia*. Brilliant red in colour,
it is also characterised by the nodular
outgrowths on its carapace and at each
of the joints on its legs.

5 COMMON SPIDER CRAB
Maja crispata
A master of camouflage, this crab lives in
most habitats and fastens bits of algae and
sometimes sponge all over its carapace, legs
and pincers. Owing to its fairly sedentary
nature and tendency to freeze when it
senses danger, it is very rarely seen as it is
often so overgrown that it resembles a lump
of seaweed-covered rock. It can grow to
over 7cm (2.75in) and is sold commercially.

6 VELVET SWIMMING CRAB
Necora puber
This swimming crab is characterised by
its blood-red, stalked eyes, and has a brown
to black-edged carapace. A very aggressive
species, it will not hesitate to defend its
territory, usually a ledge on a rock face or
a recess under a boulder. The carapace is
around 5–7.5cm (2–3in) across and it has
three blunt tubercles between the eyes. Its
hind legs are flattened and covered in hairs.
Although they are used for swimming,
they are not paddle-shaped.

7 SALLY LIGHTFOOT
Percnon gibbesi
The Sally Lightfoot is regarded as one of
the most invasive species to have ever
entered the Mediterranean. It is found on
both sides of the Atlantic, and the Pacific
coast of North America. Its carapace is only
around 3cm (1.2in) and it is very active in the
surf line amidst the rocky shorelines, where
its lively behaviour inspired its alternative
name of Nimble Spray Crab.

8 LANCER SWIMCRAB
Portunus hastatus
Commonly seen at night on fine sand or
muddy seabeds all over the Mediterranean.
The carapace is normally over 5cm (2in)
across. The back pair of legs have a dark
red blotch on the swimmerets. It is quite
aggressive in nature.

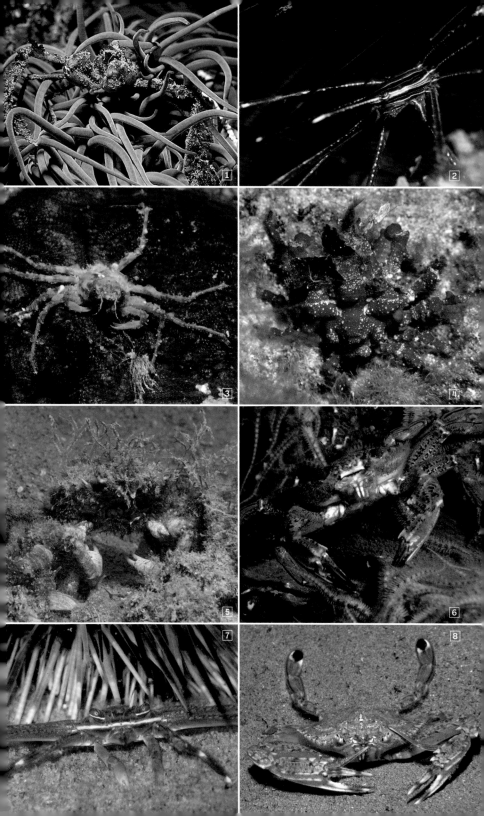

WORMS

THERE ARE MANY DIFFERENT groups of worms to be found in the Mediterranean and they are as diverse as they are interesting. There are the errant flatworms, free spirits to wander at will; nemertid worms, which are unsegmented and have eyes and a mouth and are active hunters on the reef; polychaete worms that generally produce a parchment or calcified tube, which they can hide inside when not feeding; bristle worms that look like centipedes and are active predators on the reef; and annelid worms, which burrow in soft sand and mud.

Flatworms

FLATWORMS HAVE LEAF-LIKE, flat bodies, and undulate over the seabed by the aid of tiny cilia that beat rhythmically on their underside. Carnivorous animals, their mouth and anus are located on the underside and the head has two lobed projections, which form antenna. They are distinguishable from nudibranchs by their lack of gills.

1 PINK FLATWORM
Prostheceraeus giesbrechtii
This is the most commonly seen flatworm over the entire region, primarily due to its rather garish purple-and-white striped coloration. Growing to around 2cm (0.75in) in length, it enjoys many different habitats from caves to open reefs.

2 BROWN FLATWORM
Yungia aurantiaca
Around 2cm (0.75in) in length, this flatworm has a golden to light brown body with a delicate white stripe around the outside of the body mantle or skirt. The body may also show a pattern of white spots over the entire length.

Annelid Worms

3 GREEN TONGUE
Bonellia virldis
This small species is instantly recognisable by its long, green, tongue-like proboscis, which splits into two at the end. It can extend this proboscis over 80cm (2ft 8in) from its holdfast under stones or from a rocky crevice. All mature *Bonellia* are female and are well known for being able to determine the sex of their larvae. If the drifting larvae, which land on the large proboscis, are male, they cling to the body and change into dwarf males, which then enter the intestines and sexual organs of the female. If the larvae are female they do not attach themselves but will settle onto a suitable habitat and develop into females within a year.

Polychaete Worms

POLYCHAETE WORMS ARE segmented worms with eyes and are most commonly seen as a tuft of bristles or fans, which protrude from a tube. These are the errant worms, which wander at will, and the sedentary worms, which are attached.

Errant Worms

4 BEARDED FIRE WORM
Hermodice carunculata
This species is found in all of the world's tropical and subtropical waters, being as common in the Caribbean as it is in the Mediterranean. It grows up to 30cm (12in) and is an active predator on the reef. It is characterised by the pale brown to gold or red iridescent, slender, segmented body with a series of tufts of white bristle hairs on its flanks. If accidentally touched, its needle-sharp hairs easily penetrate the skin, causing severe irritation that feels not unlike a bad burn.

1 SEA MOUSE
Aphrodita aculeata
Around 10cm (4in) long, the Sea Mouse has an elongated oval body covered in fine, grey hairs. The flanks are the most distinctive, having iridescent green and golden hairs or bristles, and lustrous brown spines. An active hunter on soft sand or mud, it is able to burrow into the substrate.

2 BOOTLACE WORM
Lineus longissimus
This is a fairly common worm living amongst stones and seaweeds, both intertidally and subtidally. Generally a rich reddish brown in colour, it can reach over 45cm (1ft 6in) long.

3 RAG WORM
Nereis diversicolor
The common Rag Worm grows to around 12.5cm (5in) long and has a rather flattened, soft body. Usually orange or brownish red in colour, it has very short tentacles and has a distinctive red blood vessel along the entire length of its back. It is found on the lower sandy shore when the tide is far out.

Sedentary Worms

4 DELICATE CORAL WORM
Filograna implexa
Filograna is one of the most delicate of all the tube worm species. Growing in large interlacing colonies over 15cm (6in) in diameter, its protective tubes are calcareous and only 1-2mm (about 0.1in) in diameter. Often growing attached to fan corals, the crown or fan tuft is only 5mm (¼in) across and its base is pale yellow or orange, giving a spotted appearance to the colony. Extremely fragile, the tubes are very vulnerable to damage.

5 MUD WORM
Myxicola infundibulum
Quite distinctive in appearance, *Myxicola* enjoys a muddy or soft sand substrate, quite often at the entrance to caverns where there is a continuous collection of detritus. This surprisingly versatile species of sedentary worm has been observed by the author around the British Isles, Scandinavia and even off Newfoundland, where it enjoys an interesting habitat around several deep wrecks at temperatures plunging to 0°C. Usually pale purple or greyish white in colour, the protruding fan is crater-like in appearance and is quickly retracted should danger threaten. Each of the tips of the fan is dark brown and the tentacles are connected by a thin film. The parchment-like tube is rectangular and rarely protrudes above ground level. The worm is 3cm (1.25in) long.

6 INQUISITIVE TUBE WORMS
Polycirrus spp.
Often overlooked by divers, these small worms are easily recognised by their trails of numerous thread-like tentacles, which snake over the rocky surface searching for food particles in the surrounding area. The main body of the worm is always hidden inside a hole or rocky crevice and is quite often additionally protected by sponge growth. The white tentacles can be over 15cm (6in) long.

7 SMALL TUBE WORM
Pomatocerous triqueter
Often overlooked due to its small size, this tiny tube worm is quite common on the lower shore and can be found on many stones and rocky surfaces. It has an obvious diamond-shaped tube, often overgrown by algae. The mouth of the tube can be tinged from green to red. The spiral fan tuft is spotted and can be virtually any colour, from white and blue to green and purple. The raised calcareous tubes are usually curled and can reach around 5cm (2in) long.

8 WHITE TUFTED WORM
Protula tubularia
Often mistaken for *Serpula vermicularis* (see page 154), this species is generally larger with the fan tuft being around 3cm (1.25in) in diameter. It is generally white or pale cream in colour and will have regular spots up each of the tentacles. It lacks a protective operculum or trap-door for its calcified tube.

1 VARIABLE TUBE WORM
Serpula vermicularis
This species is very distinctive in colour, ranging from brilliant orange to purple. It is only 2cm (0.75in) in diameter and slightly horseshoe-shaped. The protective cap or operculum of its calcified tube home is clearly visible. Enjoying rocky walls and low light in well-aerated water, this tube worm is light- and pressure-sensitive and can quickly retract its feeding fan should danger approach. Often mistaken for *Protula tubularia* (see page 152).

2 SPIRAL TUBE WORM
Spirographis spallanzani
This is the largest of the tube worms and grows out on long thin parchment tubes about 30cm (12in) long. It enjoys a mixed habitat of cave and cavern, rock wall or *Posidonia* beds. Spirograph-shaped, the fan may consist of 8 or 10 whorls, which are varicoloured. They are light- and pressure-sensitive and often withdraw back into their tubes before you can get near them. The fan may be as much as 15cm (6in) in diameter.

3 PEACOCK WORM
Sabella pavonina
A common species, generally inhabiting soft sand or muddy sea floors in sheltered low-light conditions. The tube is of parchment construction and is quite stiff. The head of the fan is around 7cm (2.75in) in diameter and is coloured in shades of brown or orange in concentric rings. It is more typically associated with the cliffs and wrecks of the Scottish west coast and the fjords of Scandinavia, where it can form large colonies. Light- and pressure-sensitive, its fans quickly retract into the tubes should danger threaten. It appears to be equally at home in strong currents or sheltered deep inside caverns.

ECHINODERMS

THIS IMPORTANT PHYLUM is entirely marine in origin and contains many different families. Completely varied in form, they nevertheless all have common characteristics, such as calcareous plates, a water-based vascular system, and tiny tube feet for locomotion and capturing food. Many are armed with some degree of protection. Most obvious of all are the sea urchins, which are covered by sharp spines.

Brittlestars

4 BLACK BRITTLESTAR
Ophiocomino nigra
Fairly common in deeper water and not always black in colour as the name would suggest, this brittlestar has five to seven rows of short spines along each arm. Its arms are up to 10cm (4in) long. It feeds on detritus and can occur in large numbers.

5 LONG-ARMED BRITTLESTAR
Ophioderma longicauda
This species has a circular disc, which is often patterned and up to 2.5cm (1in) in diameter, with arms as long as 15cm (6in). Each arm has small rows of blunt spines. This species prefers low light and is most often found deep inside caves and caverns or in very deep water.

6 FRAGILE BRITTLESTAR
Ophiothrix fragilis
This is a colourful species with the disc and arms coming in almost every colour imaginable and often varied on each individual. Hairy in appearance, the arms, which grow up to 6.3cm (2.5in) long, are characterised by having numerous rows of quite long spines. It can occur in extremely dense beds in deep water.

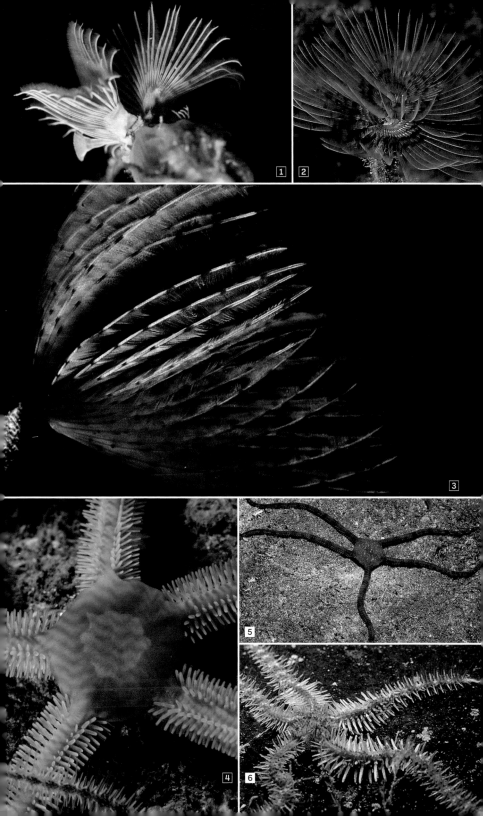

Crinoids

① FEATHER STARFISH
Antedon mediterranea
This distinctively shaped crinoid has a tiny body with the mouth on the upper surface of the cup, surrounded by 10 long flexible arms up to 10cm (4in) long, which have slender lateral branches. Under the cup-like body are a number of flexible feet or cirri with which it holds on. Usually reddish orange in colour and sometimes with banded arms, it is commonly found on wrecks and likes strong currents or well-aerated water.

② SEA STAR or BASKET STAR
Astrospartus mediterraneus
Usually found below 30m (100ft), this large basket starfish has 10 highly ramified arms. Sensitive to light, it is only seen at depth or hidden in crevices, and comes out at night when it extends its arms into the current to feed on passing plankton, forming a type of basket or net. Usually cream in colour, it is becoming more common and is found in all parts of the Mediterranean.

Starfish

③ BURROWING STARFISH
Astropecten irregularis
This is a fairly large burrowing starfish, growing up to 20cm (8in), and has five distinct arms. It is pale cream and has a uniform texture with the upper surface covered in paxillae. It has well-defined marginal plates bearing short, erect spines. It is found on fine sand and is usually partially or wholly buried. It feeds on small bivalves, worms and crustaceans.

④ IRREGULAR STARFISH
Coscinasterias tenuispina
This species can have 6–12 arms, often of different lengths, probably owing to damage inflicted whilst the starfish is young and retains the ability to regenerate limbs. It is usually light cream, mottled with brown, red and even bluish or purple blotches. Growing to around 20cm (8in) in diameter, it occurs in most habitats and feeds on the rich algae found in shallow water.

⑤ RED SEA STAR
Echinaster sepositus
This is a very common, rough-skinned, red starfish with five pointed, circular-sectioned arms up to 18cm (7in) long. The rough skin is actually papulae or gill structures and it has a slimy feel. Found in most depths, it feeds on various algae.

⑥ SPINY STARFISH
Marthasterias glacialis
One of the largest of the starfish found in these waters, it has five very long, subcylindrical arms that are roughly tapering and covered in longitudinal rows of thick, sharp spines. Each spine is surrounded by a raised group of pedicellariae. This species prefers a rocky substrate and is mottled greenish brown. It grows to over 70cm (2ft 4in) in diameter.

⑦ LONG-LIMB STARFISH
Ophidiaster ophidianus
Probably the most brilliantly coloured of the large starfish found in the Mediterranean, this species has long, cylindrical, blunt-tipped arms, which may be red (sometimes brilliantly bright), violet or orange. It can grow to over 50cm (1ft 8in) in diameter and is quite rough in texture with a slightly fuzzy appearance.

Sea Urchins

1 BLACK SEA URCHIN
Arbacia lixula
Once thought to be the male version of
the Rock Sea Urchin, the Black Sea Urchin
enjoys a well-lit, soft rock habitat where
it burrows into the rock whilst eating the
algae, which grow into the recess. It grows
to around 8cm (3.25in) in diameter and
has a black shell with black spines.

2 ROCK SEA URCHIN
Paracentrotus lividus
One of the most common sea urchins in
the western and central Mediterranean,
this species occurs in a variety of colour
forms and has a rather flattened, round test
or shell. Growing to around 7cm (2.75in) in
diameter, it burrows into soft limestone,
leaving cavities in the rock wall, which
it enlarges as it grows. It is a herbivore,
feeding on algae around its hole or over
rocky surfaces.

3 VIOLET SEA URCHIN
Sphaerechinus granularis
This is a large, globular sea urchin,
growing to around 13cm (5.25in) in
diameter and coloured a vivid purple,
usually with white-tipped spines. It likes
open, well-lit ground where it browses on
algae and has the habit of attaching small
pieces of detritus, shell fragments or
algae to its spines for camouflage.

4 LONG-SPINED SEA URCHIN
Centrostephanus longispinus
This species is relatively rare in the
Mediterranean, having been almost wiped
out a few years ago by a virulent disease.
Unmistakable due to its long, dark spines,
up to 13cm (5in) long, which are sometimes
banded, this sea urchin prefers low light
conditions and is usually seen in deep water.

5 ARROW or RED LANCE URCHIN
Stylocidaris affinis
This species is equally at home in North
Carolina, the Gulf of Mexico, Bermuda and
the Mediterranean. I photographed this
specimen off Catania in Sicily. Very easy
to identify, it has only around 15–25 very
distinctive spines which are rough in texture
and usually covered in algae or small tube
worm growth.

6 SEA POTATO
Echinocardium cordatum
The common name refers to the brittle
empty test or shell, which is sometimes
seen on the sandy seabed or shore. When
alive, it is covered in a mat of fine, cream
spines, almost like hairs. Slightly heart-
shaped, fairly round and up to 7.5cm (3in)
across, it burrows in the fine sand and is
a deposit feeder.

7 PURPLE HEART URCHIN
Spatangus purpureus
Also heart-shaped, this species can
grow as large as 12cm (4.75in). It is brilliant
purple with light, fine spines. It only burrows
shallowly and prefers coarse sand and
shell gravel.

Sea Cucumbers

[1] COTTON SPINNER

Holothuria forskåli
The most common of the sea cucumbers, this species comes in two colours – either jet black or variably tan and brown with brown spots on a cream base. Its name comes from its defence mechanism, whereby it ejects sticky, white thread when threatened. It can reach over 20cm (8in) in length and has a dense, tubercle-covered body and numerous tube feet. Feeding on sand and mud particles, which it ingests and passes through its body cavity, its excreta is usually recognised easily as it resembles an egg-like necklace chain made of fine sand particles.

[2] WHITE SPOT CUCUMBER

Holothuria polii
Also distinctively coloured, this species is black or dark brown in colour and is covered in tiny, white, spot-like spines. It grows to around 20cm (8in) and is an active detritus browser.

[3] TUBULAR CUCUMBER

Holothuria tubulosa
The longest of the commoner sea cucumbers, the Tubular Cucumber can grow over 30cm (12in) long and is varicoloured with dark brown on the upperside and light brown to peach shades on its flanks. It has conspicuous, pointed tubercles all over its body and prefers to live at the base of cliffs and amongst seagrass beds.

[4] TREE SEA CUCUMBER

Cucumaria planci
Like many other echinoderms, this sea cucumber has 10 ramified arms or tentacles around the mouth area. The body is usually creamy brown but is more often than not hidden amongst rocks or in a crevice. The dark brown or black arm tentacles are the parts most often seen; these can grow to 10cm (4in).

ASCIDIANS

ASCIDIANS, ALSO REFERRED to as tunicates or sea squirts, give the appearance of being very simple animals. However, they are very closely related to vertebrates and other chordates, possessing a complex gut and a rod-like structure similar to a spinal cord. They are either solitary or occur in large communal groups. Some even form large colonial formations.

[5] CONICAL TUNICATE

Aplidium conicum
This ball-shaped structure grows to around 20cm (8in) across and has numerous small apertures all over the body. Coloured almost translucent white to cream, it resembles a sponge.

[6] COMMON COLONIAL TUNICATE

Aplidium proliferum
Orange to red in colour, this colonial tunicate is widespread in the eastern Atlantic and middle to western Mediterranean, having migrated through the Strait of Gibraltar. At the Strait, a colony over 120cm (4ft) has been observed, making it one of the largest in the world.

1 CIONA

Ciona intestinalis

Usually found on its own, this sea squirt grows to approximately 15cm (6in). It has a white or translucent body with lobed ends to its siphons. These lobes may also be tinged with bright yellow. The inhalant opening is on the top and the exhalant is about a third of the way down on the side. In colder waters, this large species is often associated with *Ascidiella aspersa*, another similarly sized tunicate. Its range covers the whole of the north Atlantic, but it is especially abundant in the Scottish sea lochs and southern Norway. Ciona occurs from the lower shore down to 500m (1,650ft) and grows on rocks, boulder cliffs and algae. It is particularly fond of artificial structures, such as piers, shipwrecks and buoys, and it is thought that its very wide distribution is due to its propensity to attach itself to ships' hulls. It reproduces year-round, once it exceeds 2cm (0.75in) in height.

2 LIGHT BULB TUNICATE

Clavelina lepadiformis

A very similar species to the Bluestriped (below), this sea squirt grows in small attached colonies and is around 1.5cm (0.5in) tall with distinctive, white, lightbulb-filament markings within the transparent body cavity. It should not be confused with the Ball Tunicate, where the individuals are tightly packed together to form a ball shape; Light Bulb Tunicate colonies are much more loose, almost to the point of having individual zooids. These zooids are cylindrical and smooth, and you may be able to spot either amber- or red-coloured eggs or tadpole-shaped larvae within the body cavity. This species is occasionally found in deeper rock pools, but it is generally more associated with harbour walls, vertical cliffs and ledges.

3 BLUESTRIPED LIGHT BULB TUNICATE

Clavelina dellavallei

This is the largest of a distinctive small group of sea squirts. Growing alone, but in association with other individuals, it reaches around 2.5cm (1in) high. Tinged with blue lines and a yellow internal ribbing, it is unmistakable.

4 GLASS BELL TUNICATE

Clavelina nana

The smallest of the bell tunicates at only 7mm (0.25in) high, it grows in loose aggregations on algae fronds and around the base of sea fans. Its stalked, creamy white body is patterned with horizontal lines throughout.

5 DENDRODOA

Dendrodoa spp.

These small, orange or red sea squirts, only 8mm (0.25in) in size, prefer low light but well-aerated conditions and can form quite large colonial patches. They are slightly fuzzy in appearance, with the siphons next to each other. It is commonly associated with various hydroids, which grow amidst the colony. *Dendrodoa* species are also the tunicates most commonly associated with the sponge *Clathrina*, where the sponge's lacy network formations tend to weave their way amongst the small bells of the tunicate.

6 BALL TUNICATE

Diozone violacea

This develops as a tightly packed group of individuals, which resemble *Clavelina lepadiformis* in structure, but here they are grouped together to form a ball reaching 20cm (8in) in height and over 40cm (1ft 4in) diameter.

7 RED SEA SQUIRT

Halocinthya papillosa

This is the most distinctive of the ascidians and individuals are usually a deep red colour. Some individuals may be a pale pink in deeper caves, but the siphons are still tinged dark red. It prefers a solitary existence and grows to around 12cm (4.75in).

PLANTS

SEAGRASSES

SEAGRASSES ARE THE ONLY flowering plants in the sea and there are three representatives to be found in the Mediterranean. The flowers are tiny and obscure, appearing in the spring and early summer. Seagrasses anchor the seabed and spread by their root-like rhizomes.

1 CYMODOCEA
Cymodocea nodosa
Less common than Neptune Grass, the leaves of Cymodocea are grey-green in colour, reach only 50cm (1ft 8in) long, and have lacy edges, joined in tufts at the base. They grow from a thin rhizome. This plant prefers warm, calm waters and a fine surface sediment. When in flower, its blooms appear at the tip of the stem, but are rarely seen.

2 HALOPHILA
Halophila stipulacea
This is a Lessepsian migrant from the Red Sea and colonies have reached as far west as the islands of Malta, Gozo and Sicily. It has a rounded top to the 4cm (1.5in) green blade. Patches spread by a common root system with each new plant forming around four to six leaves. Other than its invasive nature into other areas of the Mediterranean, indications are that it does not compete with other Mediterranean seagrasses. But ultimately, all migrants are competitive and studies are still under way.

3 NEPTUNE GRASS
Posidonia oceanica
This is by far the most abundant seagrass found around the Mediterranean shoreline. Growing on sand and detrital bottoms, it forms huge meadows in localised areas and is an indicator of clear, clean water. Growing very densely, the leaves are up to 1m (3ft 3in) long and are an important habitat for all manner of marine life. The University of the Algarve has now done extensive DNA tests on *Posidonia* and has discovered that some of the meadows are more than 200,000 years old, dating them to the early Pleistocene and the 'Dawn of Humanity'.

The detritus from *Posidonia* is often washed up on wind-swept beaches, and forms small seagrass balls (see inset photo). The University of Barcelona has shown that these 'Neptune Balls' also contain plastic debris which collects at the base of the plants. The University's studies show the vital role that seagrass meadows have as they absorb carbon dioxide and release oxygen, as well as providing the perfect natural hatchery for many marine species. Seagrasses like *Posidonia* are the foundation of all the coastal food webs.

ALGAE

ALGAE ARE INCREDIBLY WIDESPREAD and diverse, with several hundred species being recorded in the Mediterranean, and more being discovered each year as migrants from other, more distant seas. Some are single-celled, living in the tissue of some hard coral polyps, while others are multicellular, spreading over rocky surfaces. Many grow in tufts and some are so delicate that it is hard to imagine that they are, in fact, algae. Many species are eaten by particular species of herbivores, whether they be fish, sea urchins, snails or others. Algae can be split into three groups depending on their pigmentation. These groups are the green, red and brown algae, most of which have relatives in other oceans of the world.

Green Algae

1 CHLOROPHYTA or MERMAID'S CUP
Acetabularia mediterranea
This unmistakable alga consists of clumps of thin green or white stalks topped with a thallus of a light green, rayed disc hardened by encrusted limestone. Umbrella-shaped, the stalks grow around 8cm (3in) tall with the discs being around 1.2cm (0.5in) across. In the growing season, it is eaten by various wrasse and bream species, which prefer the discs, leaving many empty stalks. Large clumps are concentrated on a stone or rocky substrate. It is quite often found on the lower shore and can grow as deep as 30m (100ft).

2 COMMON CAULERPA
Caulerpa prolifera
This indigenous species of *Caulerpa* is still fairly common around the Mediterranean, particularly in southern and central areas where its cousin *C. taxifolia* (see opposite) is not as rampant as it is in the north. The two species have similar ways of spreading over all substrata by way of a common robust stolon or stalk. Common Caulerpa has wide branching blades protruding perpendicularly into the water. The blades reach at least 15cm (6in) long and can be as broad as 13cm (5in). This alga is also very common in the Caribbean and is one of the species favoured by marine aquarists.

3 CREEPING CAULERPA
Caulerpa racemosa var. occidentalis
This is a highly invasive species, originally thought to have come through the Suez Canal from the Red Sea. However, indications are that the species is actually Australian in origin. Interestingly, *C. racemosa* is very prolific in the Caribbean and is located as far north in the Atlantic as Bermuda. It is also common in the Pacific and is used as a food source by Polynesians, who eat it raw with grated coconut and coconut milk. This alga spreads by means of long stolons. Its racemous rhyzoids point downwards and photosynthesis occurs in the clustered aggregations, which reach around 5cm (2in) in height. Bright green, with rounded lobes to each branch, this species appears to be both warm- and cold-water tolerant, occurring from 5m (17ft) to below 50m (165ft) depths.

4 CARPET ALGA
Caulerpa taxifolia
First introduced accidentally in 1984 in Monaco, this virulent alga has leaves that are small and paired, with the new growth being a pale green or even yellow. With no natural enemies in the region, it is growing at such an alarming rate that the entire Mediterranean is at risk. It spreads from a single stalk, and an enzyme it produces is toxic. Leaves can be from 6–65cm (2.5in–2ft 2in) long, and in 1m² (11ft²) of seabed an individual plant can have a stolon of more than 3m (10ft) in length, with more than 200 leaf fronds. In total, more than 220m (720ft) of plant material with 8,000 leafy fronds, weighing 12kg (26lb), make up 1m² of new Carpet Alga meadow. *Caulerpa taxifolia* is very similar to *Caulerpa sertularoides*, which is often used in salads in the Philippines and is not in the least bit toxic.

1 FINGER CODIUM

Codium vermilara

Noted for its small clumps of 'fuzzy' fingers, this alga has a common root stalk and branches widely at its base to form many small rounded clumps of stalks. Quite often coated in a filamentous, slimy alga, Finger Codium is grazed on by a few species of nudibranch. This species enjoys aerated, shallow water conditions where wave surge and high light are present.

2 PURSE CODIUM

Codium bursa

Unmistakably rounded shape in various shades of green. Growing up to 40cm (1ft 4in) in diameter, as it gets older the ball develops a central depression. Found singularly or in small groups of various sizes on a rocky substrate, the alga's exterior can get rather scruffy in appearance and quite often has other species of algae adhered to its surface.

3 SEA CACTUS

Halimeda tuna

Very similar in formation to *Halimeda* species found in the Caribbean, the Sea Cactus is made up of a series of distinct small rounded discs approximately 10cm (4in) long, which are joined together by a narrow strand. Coloured from a pale brownish green to a brilliant green of new growth, the plant resembles the terrestrial 'Prickly Pear' cactus. Often found under overhangs, it prefers shaded or deeper water conditions.

4 SEA FAN

Udotea petiolata

Typically fan-shaped with a convoluted, rounded edge, attached to rocky surfaces by a short peduncle. Quite often, there are darker green growth lines over the leaf which can grow up to 10cm (4in) high. During reproduction, the outer edge is tinged white. Growing in clumps and widely spread out over rocks, this alga prefers shallow, light-filled, warm water.

Red Algae

5 RHODOPHYTA or ROUGH CORAL MOSS

Corallina elongata

A red or pale pink seaweed made up of calcified parts loosely connected into a feather-like shape. Whilst growing, the new growth may appear lighter in colour than the stem. The tips of the 'feathers' are quite often all that you may see as it grows through other species of algae and small sponges. However, it can also grow to form quite dense carpets of algae. Preferring shallow water, the fronds are 2–6cm (0.75–2.5in) long. It can resist rough seas and moderate pollution. This alga, when dried, was formerly used medicinally with other red algae for the treatment of intestinal worms.

6 JANIA

Jania rubens

Similar in structure to *Corallina elongata*, but much paler and thinner, Jania comprises a large clump of calcareous and jointed fronds, which branch off dichotomously. The reproductive organs are in the form of a swelling at the joints and difficult to see. Growing from 1.5–4cm (0.5–1.5in) in diameter, the rosy red alga forms circular clumps in shallow water and is quite often associated with numerous other algae which it tends to colonise.

7 STONE WEED

Lithophyllum lichonoides

Sometimes referred to as 'footpath alga' or 'trattoirs', Stone Weed is strongly calcified and is common in surface waters where it forms a rim or cornice around shoreline rocks. The growth is determined by tidal movement and the cornices can be quite wide. Wherever it occurs in deeper water, it can be found on rocky bottoms in areas of strong tidal movement. Coral-like, with numerous hard and flattened fronds, it is purple in colour with white or pink new growth lines at the edges. It is very tolerant of wave action, preferring shallow water where it can become exposed at low tide. Similar in shape to *Pseudolithophyllum expansum* (see page 170), as the individual parts of this alga grow, they fuse with other parts to form large reef-like structures.

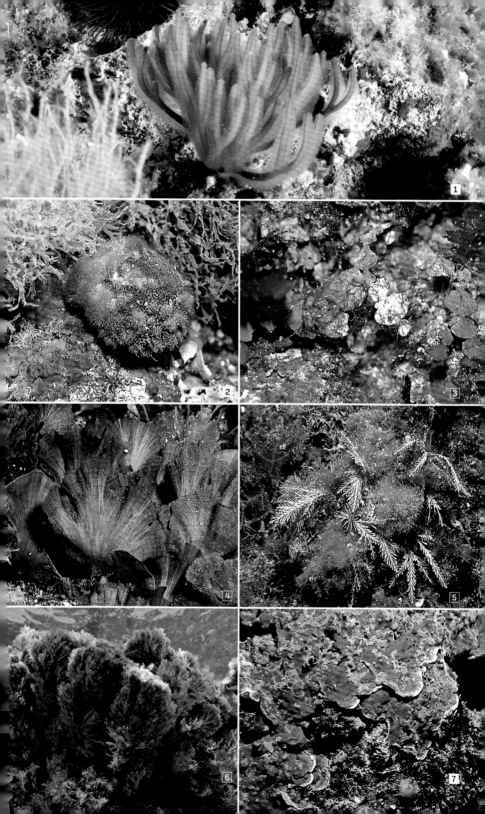

1 SEA ROSE
Peyssonnelia squamaria
Bright and colourful, the Sea Rose is brownish red in colour and each part forms a small fan, progressively getting larger as the fan grows out in petal-like formations. Each 'petal' is pale yellow at the base, with striated lines that grow stronger in colour as they expand to the outer rim. Often appearing almost luminescent underwater, each rose may grow to 10cm (4in) in diameter and they are often associated with a brilliant green alga. This colourful species prefers low light conditions, either at the entrances to caves and caverns or on poorly illuminated hard rock bottoms where its range can extend to below 50m (165ft).

2 SEA MAT
Pseudolithophyllum expansum
Sea Mat is quite distinctive in shape and colour form, being a pale purplish pink with a light outer rim. This very hard coralline alga encrusts large areas of poor light and is found on most rocky substrates from shallow waters to below 60m (200ft). The growing edge of the alga can be deeply convoluted and is one of the contributory algae that forms the Mediterranean reefs, cementing detritus, algae and rock together.

3 RED SEAWEED
Sphaerococcus coronpifolius
This species is particularly common in most coastal regions of the Mediterranean. Dark ruby red in colour and forming small clumps widely branching from a single stalk, this seaweed prefers well-aerated, light-filled, shallow water. Quite often, wherever there has been damage to the rock surface by anchor or boat groundings, this alga is one of the first to take hold. Extremely opportunistic, this quick-growing alga may dominate entire rocky surface areas.

Brown Algae

4 PHAEOPHYTA or FORKED RIBBONS
Dictyota dichotoma
This is a delicate alga, the green chlorophyll masked by brown pigments, changing to a brilliant iridescent blue at the outer edge of its tips. It usually only grows to around 15cm (6in) high, with dichotomously branching, ribbon-like fronds 2–12mm (0.1–0.5in) wide. With no mid-rib to support the alga, it wafts in whatever current assails it. Fragile in nature, it is easily broken apart or knocked off its base. Quite often found washed up on the shore.

5 SEA FERN
Halopteris scoporici
Widespread on all well-illuminated rock surfaces in central and eastern Mediterranean regions, the Sea Fern grows to around 15–20cm (6–8in) tall and has a thick, dark brown (almost black) single stalk, branching to dense feather-like tufts which are rough in texture. Found in all depths, it is often overgrown by other species.

6 PEACOCK'S TAIL
Padina pavonica
Calcareous in nature, the fronds have a parchment-like feel to them. Each frond is flat and fan-shaped, changing to become funnel-like as it matures. Coloured light brown to white, it has darker, horizontal, striped growth lines. Tolerant of some freshwater run-off, it is quite often found on jetty slipways and pilings. Enjoying sunlight and warm calm waters, it is rarely found below 20m (66ft) and has arrived via the Red Sea and Suez Canal.

7 COMMON SARGASSUM
Sargassum vulgare
Sargassum weed can grow up to 1m (3ft 3in) in height and is widely distributed around the Mediterranean. It is fast-growing and attaches to a rocky substrate. It has long, crinkly fronds with short, lateral, leaf-like structure branches, interspersed with small air bladders. It will tolerate some pollution and if broken free from its rocky base, the alga survives quite happily on surface waters, buoyed by its air bladders. It provides a home and shelter for many small pelagic creatures.

GLOSSARY

Aggregation: Individuals gathered together in a loose group. Can apply to fish or to attached (sessile) animals.

Anal fin: Single fin running along the underside of a fish, from the anal opening towards the tail.

Antennae: Projecting feelers, which have a sensory function.

Appendage: Parts of an organism that project from the body and have a specific function.

Benthic/Benthos: Of the seabed, and animals and plants that live there.

Carapace: The hard upper or outer shell of a turtle or crustacean.

Caudal: As in caudal fin – a fish's tail.

Cerata: The fleshy extensions or tentacles on the back of a nudibranch, which often store toxins from its prey to help in its own defence.

Chelipeds: A crustacean's claws.

Circumtropical: Found in all tropical oceans of the world.

Colony/colonial: A group of animals living together, or an apparently single organism that consists of numerous similar or identical interdependent 'units' (for example, polyps) that are linked together to form the complete organism.

Commensal: Relationship in which two species live in close physical association, with one sometimes dependent on the other for some function or other.

Coralline: Resembles the material from calcareous coral skeletons.

Detritus: Particulate matter of algal and animal origin, which can become attached to the spines of various crustaceans and fish.

Dorsal: The upperside of an organism – see also Ventral.

Dorsal fin: The fin on the top of the back in fish and cetaceans. It may be single or split into two or three separate sections.

Ectoparasite: Animal that lives on the outside of another, and is dependent on it for its nutrition (for example, it attaches to the skin and feeds on blood).

Herbivorous: Animals that feed on plants, such as algae and seagrasses.

Hermaphrodite: Individual that has both ovaries and testes. These may be functional at the same time, or the animal may function first as one sex and then the other.

Invertebrates: Animals without backbones.

Lessepsian: All species of marine life that have travelled through to the Mediterranean from the Red Sea are known as Lessepsian migrants, after the Suez Canal engineer Ferdinand De Lesseps.

Mantle: The fold of skin which extends from the body of a mollusc such as a cowrie, and may completely cover the outer shell.

Medusa: The free-swimming stage of jellyfish.

Nematocyst: The harpoon-like, stinging spike which certain invertebrates can fire, in defence or to stun prey for food.

Operculum: The hard bony plate which acts as a protective cover to the gills of a bony fish; also the protective cover over a mollusc's opening in its shell.

Pantropical: Found between the Tropics of Cancer and Capricorn around the world.

Paxillae: Small umbrella-shaped structures on the upper surface of sea urchins and starfish, which form an external protective skin.

Pectoral: As in pectoral fin, found on the sides of a fish, usually behind the gills – analogous to the front limbs of an amphibian or other tetrapod.

Pelagic: Living and swimming in the water column, rather than close to or on the seabed.

Pelvic fins: Pair of fins located on the underside of a fish, usually behind the pectoral fins.

Plankton: Plants (phytoplankton) or animals (zooplankton) that float in the water column and are moved by ocean currents. Mainly microscopic, but includes larger organisms, such as jellyfish larvae.

Polyp: Either six- or eight-armed, anemone-shaped invertebrate organism, typically sessile (used to describe representatives of several unrelated groups).

Predator: Animal that hunts and feeds on other animals.

Radula: The grinding plates in certain fish and molluscs; used in part of the food breakdown process.

Rhinophores: The sensory appendages on the head of nudibranchs.

Scutes: Horny plates that form the shell of turtles.

Spat: Baby seed stock of oysters and mussels from outside the Mediterranean.

Spicules: Calcium spikes found in sponges and soft corals.

Symbiotic: Relationship in which two (or more) species live together and benefit each other in one or more ways.

Territory: A living, feeding or breeding area that is actively defended from intruders by the inhabitant(s).

Ventral: The underside of an organism (see also Dorsal).

ACKNOWLEDGEMENTS

Over the years collecting the enclosed photographs, I have dived with many different groups, dive boats and individuals. It is impossible to thank everyone, so I give you a collective 'Many Thanks', but some deserve a special inclusion here: Evasioni Blue Diving in Ustica; Gibraltar Sub-Aqua Club; Diamond Diving in Golfe Juan; Xaloc Diving in Estartit; St Andrews Divers Cove in Gozo; Maltaqua on Malta; Alexandra Valletta from VisitMalta; The Diving Sea Club in Catania, Sicily; Proteus Diving in Sardinia; West Coast Divers in Mallorca; Pantxao in Villefranche sur Mer; Cydive in Cyprus; Posjdon Diving in Croatia; Professor Alexandre Meinesz and the University of Nice; Air Malta; Air Italia. I must pay tribute to my wife Lesley, who supports me in all my endeavours, even when it means that I have to be away from home researching yet another book project. Thank you, I really could not do this without you.

PICTURE CREDITS

All photographs and figures are copyright © the author, except for those listed below.
Bloomsbury Publishing would like to thank those listed below for providing illustrations and for permission to reproduce copyright material within this book. While every effort has been made to trace and acknowledge copyright holders, we would like to apologise for any errors or omissions, and invite readers to inform us so that corrections can be made in any future editions.

1 José Francisco Martín Piñatel/Alamy Stock Photo; 2 BIOSPHOTO/Alamy Stock Photo; 9 Dennis Cox/ Alamy Stock Photo; 10 (top) mairu10/Shutterstock; 11 and 126 (5) imageBROKER.com GmbH & Co. KG/ Alamy Stock Photo; 28 and 94 (1) Nature Picture Library/Alamy Stock Photo; 37 (1) Charles Hood; 39 (5) Luke Atkinson; 41 (2) Dan Bolt; 43 (1) BIOSPHOTO/Alamy Stock Photo; 47 (3) Blue Planet Archive/Alamy Stock Photo, (5) David Salvatori/VWPics/Alamy Stock Photo; 57 (1) Humberto Ramirez/Getty Images, (3) WaterFrame/Alamy Stock Photo, (5) BIOSPHOTO/Alamy Slock Photo; 59 (2) WaterFrame/Alamy Stock Photo; 71 Kirsty Andrews (1), Dan Bolt (4); 73 (2) imageBROKER/Rolf von Riedmatten/Getty Images, (3) Mark Doherty/Shutterstock; 75 (3) Luke Atkinson, (6) Dan Bolt; 89 (2) University of Nice; 91 (4) Kirsty Andrews; 97 (1) Shane Wasik; 99 (3) WaterFrame/Alamy Stock Photo; 109 (4) Poelzer Wolfgang/Alamy Stock Photo; 123 (8) WaterFrame/Alamy Stock Photo; 133 (5) Bruno Manunza/Alamy Stock Photo; 135 (3) Avalon.red/ Alamy Stock Photo; 145 (5) Joao Pedro Silva/Getty Images; 147 (1) Bruno Manunza/Alamy Stock Photo, (6) Alex Mustard/2020VISION/naturepl.com, (7) WaterFrame/Alamy Stock Photo; 149 (3 and 5) WaterFrame/Alamy Stock Photo; 156 (2) WaterFrame/Alamy Stock Photo; 161 (5) Poelzer Wolfgang/ Alamy Stock Photo; 165 (1) Dimitris Poursanidis/naturepl.com, (2) imageBROKER.com GmbH & Co. KG/ Alamy Stock Photo; 167 (2) Wolfgang Pölzer/Alamy Stock Photo; 169 (1) Jaime Franch Wildlife Photo/ Alamy Stock Photo, Minden Pictures/Alamy Stock Photo (6); 171 (4) Damsea/Shutterstock

INDEX